Supplementary Volume no. 14

PASTORAL ECONOMIES IN CLASSICAL ANTIQUITY

Edited by

C. R. WHITTAKER

THE CAMBRIDGE PHILOLOGICAL SOCIETY
1988

ISBN 0 906014 13 1

Printed in Great Britain by the University Press, Cambridge.

CONTENTS

INTRODUCTION

Khazanov's extraordinary study of *Nomads and the outside world* opens with a warning that those who try to create a typology of pastoralism must 'be prepared not for dithyrambs but for some basic and severe criticism.'[1] Small wonder, therefore, that few have cared to do so. The papers in this volume, which were discussed during the IXth International Economic History Congress at Berne in August, 1986, demonstrate and go some small way toward satisfying the need for such an attempt.

Pastoralism is an ideal type that has never existed in pure form, not even among Masai or Berbers.[2] That is to say, there is always in every agrarian community a mixture of agriculture and animal breeding and always a spectrum in the relative importance of one towards the other. It is banal, therefore, merely to identify the existence of 'elevage' or animal husbandry. The questions that have preoccupied almost all of the contributors have been those of definition – at what point should animal husbandry be called 'pastoralism'; of identification – how can references to animals in our various ancient sources distinguish between husbandry and pastoralism; and of historical evaluation – under what conditions does pastoralism assume a predominance within the community.

Not surprisingly the answers to these problems vary considerably from one contributor to another, not least the question of definition. Cherry takes it as self-evident that 'specialized pastoralism' is the proper subject for discussion. But Hodkinson and several others wish to evaluate the whole range of pastoral activities, from animal husbandry on the farm to what Corbier calls 'true pastoralism'. Goudineau regards pastoralism as quite separate from 'élevage' and 'indissolubly' tied to transhumance and to ovicaprids. Gabba follows along this path, although he would wish to include within the definition what he calls 'silvo-pastorale arcaico' and cattle. Garnsey expresses irritation with those who refuse to concede the name of pastoralism to anything except when agriculture is negligible, but he, too, implicitly accepts transhumance as the defining theme. Even transhumance and nomadism, however, need their own typology, as Braudel and Khazanov have stressed,[3] since there are political and economic implications in the various modes of practice which modify the weight we should attach to the empirical evidence.

Leveau is right, of course, that such debates can take an 'irritating turn' towards a war of words. The reason for typologies, however, is not simply to have what Leach calls 'butterfly collections' but to avoid naive comparisons and the dangers of reductionism. The aim is to set out systematically the conditions of food production and resources in relation to population and labour (see Jongman). Even if no very clear or agreed definition of pastoralism finally emerges, all the contributors show an awareness of the relationship of animal culture to agriculture in an historical context and of the different consequences of on-farm pasturing, off-farm animal herding and long-distance animal movement. Harsh words are rightly said about former, sweeping assumptions of environmental determinism and continuity without reference to the context. Ideology has played havoc with the early histories of Greece and Rome because the implications of pastoralism have not been fully considered, as Cherry and Ampolo point out. Pastoralism, says Goudineau, serves too often as an alibi for unknown historical forces.

Typology, therefore, is naturally linked closely to methodology and evidence. Cherry's paper provides a valuable history of archaeological misinterpretations, through lack of 'a well-grounded theory' (Hodkinson). Mountain huts, shrines and sheep-shears do not inevitably add up to transhumance or specialized pastoralism. But on the other hand they might. Garnsey's criticism of Pasquinucci in this respect earned a vigorous response from her (unfortunately not recorded here). But his scepticism finds support from Goudineau and Frei-Stolba, who wrestle with the archaeological evidence from Roman Gaul and Switzerland but see no easy way of separating animal husbandry from transhumant pastoralism on the basis of the evidence alone, nor how to arrive at quantitative results. It is only by a full appreciation of the balance between pasture land and agriculture that such ambiguous material can begin to be assessed.

Even the literary evidence can lead to strikingly different interpretations. The most explicit examples here are the papers by Hodkinson and Skydsgaard. The former arrives at the opinion that animal husbandry in Greece was in general on-farm and symbiotic with agriculture, rarely involving long-distant transhumance. Skydsgaard, however, from the same evidence, believes that animal breeding was separated from agriculture and non-sedentary, although agreeing that long-distance movement was unlikely. The debate turns on the availability of fodder, with Hodkinson relying heavily on the evidence of the fourth-century polymath Theophrastus but rejecting that of Homer, while Skydsgaard does precisely the opposite. Readers must judge for themselves. In Skydsgaard's support Jameson accepts the value of Homer and concludes that in Athens there was insufficient fodder for the estimated size of herds; but he also argues that Athens was an exception because of her riches and imported grain. Similar controversy can be found over the scale and date of transhumance in Italy in the contributions from Garnsey, Thompson, Corbier and Gabba. While all are agreed that Roman imperialism and political control of Italy led to a pastoral 'industry' after the second century B.C., there is a quite sharp difference of opinion over the extent of specialized herding and seasonal movement of animals before that date. Garnsey relies heavily on the Braudelian dictum that transhumance 'presupposes complicated internal and external structures and weighty institutions.'[4] Gabba argues from anthropological studies that complementary interregional economies were feasible before the overarching political power of Rome.

Nomadism is less obviously highlighted in these papers, since only Leveau discusses the problem in the context of North Africa (although Corbier acknowledges its existence on the margins of the Roman empire). The debate, however, is no less intense between those who – like Leveau here – believe that the literary and epigraphic evidence proves that Roman authorities sedentarized and 'contained' nomadic herdsmen, and others, cited in his notes, who argue that imperial frontiers were expressedly designed to permit movements of transhumant and semi-nomadic populations and to provide for symbiotic exchange between herdsmen and farmers (with which Leveau seems also to agree).

There is no cause to be pessimistic about such controversies. Debate clarifies the issues and demonstrates where research is needed. Ancient literary and epigraphic sources are not enough. New techniques of geo-morphology (Leveau), ethnoarchaeology (Cherry), etc. are constantly eliminating guesswork. Jameson, Goudineau and Bökönyi provide examples of how much can be done with bone evidence to calculate the scale, strategy and continuity of animal breeding. The way forward obviously lies in the close cooperation of archaeologists, geographers and historians.

I draw three basic conclusions from these papers, which should perhaps serve as propositions for future research:

(1) Pastoralism must always start from agriculture. Historically, as Braudel argued, it was the farmers' determination to plough that 'opened up the way for shepherds.'[5] Ampolo, Cherry and Skydsgaard make it clear enough that the Varronian tripartite, linear evolution of agriculture from pastoralism is quite simply an 'heroic myth'. Economically, mountains or deserts are one world with the plains, not two. Conflict or dissidence over resources does not alter the fact that farmers and herdsmen need each other and, as Adam Smith saw, the price of grain and the price of fodder are intimately related.[6] Importation of cheap grain to Athens (Jameson) or to Rome (Gabba), just as into Holland in the seventeenth century, made herding more profitable than ploughing. At a social and demographic level many have argued that rising populations have often forced specialized pastoralism or sedentary animal husbandry.[7]

(2) Specialized pastoralism must be accompanied by specific political conditions. The danger of ecological reductionism, to which all contributors are sensible, implies that political superstructures must reflect the economic base.[8] The Spanish 'cañadas', the 'drailles' of Languedoc, the 'traturri' of Italy functioned under the guarantee of institutions like the Castilian princes, the Knights Templar or the Roman emperors, who charged a price for their

protection. But were these the only political conditions possible for long distance transhumance? Aragón also practised transhumance but without the political institutions of neighbouring Castile and not without conflict.[9] Gabba and Skydsgaard suggest that terms of reciprocal agreements could also facilitate movement of animals without political domination.

(3) 'An economy founded on pastoralism is not infrequently an economy of relatively high involvement in the market.'[10] The principle is invoked by a number of contributors and implied by others. Gabba notes the purchase of Cisalpine cattle and Bökönyi the importance of improved stock breeding. One aspect of the latter, to which perhaps insufficient attention has been paid in these papers, is the needs of the army, whose demands for cattle products and wool in the Roman Empire was enormous. Was it this that stimulated the villas of the Mittelland of Switzerland (Frei-Stolba)? Jongman refers to space for cattle herding beyond the German frontiers, which certainly served the military market.

The longer papers in this volume are more or less the communications delivered at Berne. A number of shorter communications were made by participants and three of these have been included at the end of the main papers.

It is not out of place here to thank Professor Herzig and Dr Frei-Stolba of the University at Berne for all their help and hospitality during the Congress.

Churchill College, Cambridge C.R. WHITTAKER

NOTES

1. A. M. Khazanov, *Nomads and the outside world* (trans. 1984) 5.

2. W. Goldschmit in *Pastoral production and society* (1979) 16.

3. F. Braudel, *The Mediterranean and the Mediterranean world in the age of Philip II* (1975) I 86-7; Khazanov (n.1) 19-21.

4. Braudel (n.3) I 94.

5. Braudel (n.3) I 179.

6. A. Smith, *The wealth of nations* (ed. E. Cannan, 1976) 168.

7. E.g. X. de Planhol in *Pastoral production and society* (1979) 29-42.

8. G. Dahl in *Pastoral production and society* (1979) 262.

9. M. J. Walker in *Transhumance and pastoralism* (*World archaeology* Spec. issue 15 (1) ed. W. Bray) 37.

10. Khazanov (n.1) 202; K. wrote 'pastoral nomadism' but I do not think my alteration does violence to his sense.

PASTORALISM AND THE ROLE OF ANIMALS IN THE PRE- AND PROTOHISTORIC ECONOMIES OF THE AEGEAN

Introduction

Evidence of domesticated animals in the Aegean reaches back to the 7th millennium B.C. Ovicaprids, pigs, cattle and dogs (in that order of importance), together with various wild species, comprise a recurrent economic 'package' on the first farming sites, whether in Crete (Jarman and Jarman 1968), in southern Greece (Payne 1975), or on the plains of Thessaly (Halstead 1984). With the later addition of equids and a number of other economically less important animals, these species persist throughout later prehistory down into historically recorded times as the essential ingredients of animal husbandry in Greece. One might suppose, therefore, that the study of ovicaprid-dominated animal economies in later prehistory and protohistory would be an analysis of the obvious, offering far less of interest than the investigation of the origins of this set of well-adapted man/animal relationships. In practice, however, it opens up a fascinating area for investigation, particularly as regards long-term trends in the impact of growing sociopolitical complexity, settlement nucleation and urbanisation on the organisation of animal management systems, and *vice versa*.

Domestic animal communities can be exploited for a variety of resources, among the most important being meat, milk and milk products, wool, hides, traction/transport, and manure; they can also function as protection, as herding or hunting aids, as sources of entertainment, as prestige items, as wealth that can either be stored or put into circulation at high rates of return and with considerable value in social transactions, and so on. These goals to some extent conflict with each other. The limitations of choice in specific sets of circumstances are reflected in the adoption of *management strategies* aimed at maximising economic return, or minimising the expenditure of labour, or reducing the risk of subsistence crisis, or – very often – some combination which goes some way toward satisfying all these aims. The chief means by which a given strategy is achieved are (1) by opting for particular combinations of species, and (2) by manipulating the age and sex structures of each species. The characteristic patterns that can be expected to be the outcome of differing strategies of animal management provide the archaeozoologist with interpretative models against which to measure excavated faunal assemblages. For instance, in a now-classic paper based on personal observation of sheep and goat herds in Turkey, Payne

(1973) outlined three quite distinct mortality patterns associated with herding systems, each aimed at optimising the potential for meat, wool or milk; similar models have been proposed by Higham (1967) for cattle. In practice, of course, herding can serve a mixture of purposes (Cribb 1985), and ecological constraints may also lead to significant differences between the archaeological data and such idealised models. Moreover, the problems and opportunities of keeping herds must be seen in the context of the *overall* subsistence economy, which rested squarely on the foundations of sedentary arable farming: most of the arguments for this view, set out by Hodkinson (this volume) in the case of the Greek *polis*, apply with equal force in prehistoric times. When one adds further complications, such as transformations of the natural landscape as a result of climatic change and/or human interference, or the problems associated with the formation, preservation, and archaeological sampling of animal bone assemblages, it can be appreciated that the study of man-animal relationships throughout the several millennia of Greek prehistory is neither straightforward nor devoid of interest.

I have deliberately emphasised archaeozoology – the study of excavated animal bones – at the outset, since, in the absence of written sources, it is naturally the mute evidence of the bones that must serve as the primary source of information about Greek animal husbandry in prehistory. Here, perhaps not surprisingly, the prehistoric archaeologist scores heavily over his classical counterpart. Payne (1985) has recently published a useful and thorough review of all the published information available on bone samples from sites in Greece. This makes it clear that, while bone reports exist for some 40 sites of prehistoric date, the same can be said of only 10 Iron Age or later sites (and only two or three of these are detailed studies of large assemblages). As Payne notes, animal bones are not at all rare on most sites in Greece, including those of the Iron Age and later periods: 'what *has* been rare is the classical archaeologist prepared to expend time and resources on their systematic collection' (1985:211). Consequently, despite considerable differences in publication standards and inevitable gaps in coverage, the number and size of bone samples of Neolithic and Bronze Age date do allow comparison and synthesis, on a regional and chronological basis, in a way that is currently impossible for classical antiquity.

It is not the intention here to attempt a general consideration of the role of animal husbandry in the overall economy: that topic is too large, nor is the present author qualified to tackle it. The best and most recent statements are by Halstead (1981; 1984; 1985; 1987), who summarizes a wide range of empirical data, as well as discussing alternative model subsistence strategies. The focus here, instead, is on whether there exists evidence, in the period from the origins of domestication in Greece until the end of the Dark Age, for *specialised pastoralism*. I am taking it as read (since Halstead and others have laid out the evidence very clearly) that during this period, as certainly later, Greek farming systems involved a very close symbiosis of crop production and animal husbandry. Plough agriculture for cereal and pulse production depended on animal traction power; manure – deposited either by animals while grazing stubble or fallow fields, or in certain circumstances by its deliberate transport from nearby settlements – was essential for maintaining soil fertility. Conversely, animals served to convert otherwise unusable kitchen waste, agricultural by-products such as prunings and pressings, and non-arable pasture into edible and storable meat, a resource of great importance as a hedge against the inherent risks of climatically-induced subsistence crisis. To speak of the role of animals in this kind of small-scale mixed farming system as 'pastoralism' – as, for instance, do Chang and Koster (1986:98-99) – involves an unduly catholic and potentially confusing definition of the term: on this basis, virtually all Greek subsistence farming, past and present, would involve pastoralism.

The two basic variables which many others (e.g. Irons and Dyson-Hudson 1972; Spooner 1973; Dyson-Hudson and Dyson-Hudson 1980) have used to define pastoral economic systems are a high degree of special dependence on herd animals, and (usually) an element of mobility in their exploitation. However, specialisation is obviously a relative matter, and it need not imply the large-scale pasturage of livestock away from cultivated land or the seasonal transhumant movements to and from mountain pastures that have generally been regarded as typifying animal husbandry in Greece. Hodkinson (this volume) presents plentiful evidence from ancient Greece for the intensive management, solely on smallholdings, of quite modest numbers of livestock, kept specifically for the sale of their superior quality products (e.g. cheese or lambs) to urban markets at above-average prices. This might not be termed 'pastoralism', but it is certainly one form of *specialised* livestock production. In considering the evidence from early Greece for specialised animal economies, my emphasis will not be solely on their scale, or degree of mobility, or even on the extent to which humans were dependent on animals (e.g. as providing x% of annual calorific requirements) – although it is not denied that all these are relevant aspects. I prefer instead to stress the 'differentiation and coordination of kinds of labour and other factors in production associated with herding which forms a restricted occupational group' (Galvin 1987:120). Put another way, 'the more herding can be considered an occupational specialisation, the more likely herd management practices will reflect production for exchange, rather than subsistence' (*ibid.*). This, of course, has been the central element in discussions of the origins of specialised nomadic pastoralism in the ancient Near East, the separation of herding and farming being seen as causally related to the emergence of centralised political-economic institutions mediating the exchange of their respective products (e.g. Lees and Bates 1974).

This paper first reviews some of the types of evidence which have been taken as indications of the existence of pastoral economies in the prehistoric Aegean, and then proceeds to evaluate the widely accepted postulates that some form of transhumant pastoralism is a 'natural response' to the Greek landscape and one that has therefore existed there ever since the introduction of domesticated animals. I shall argue that specialised pastoralism is an unlikely mode of subsistence in the context of the types of environmental settings and farming systems that existed in early prehistoric Greece and, indeed, is an adaptation that has emerged historically only in specific socio-economic and political circumstances. Further difficulties are posed by the general lack of theory linking pastoralism as a socio-economic adaptation to particular aspects of material culture, a problem for whose solution ethnoarchaeology appears to offer the best prospects. These general arguments serve to structure a discussion of the rôle of animals at three very different stages: early village farming in the Neolithic, the Mycenaean palace economies, and the centuries of the Dark Age. For the latter period, in particular, an emphasis on pastoralism in most archaeological and historical accounts seems to stem less from any available evidence, which is admittedly very limited, than from a widespread ideology of the pastoral nomad.

Archaeological models positing the existence of pastoralism in prehistoric Greece

Vickery (1936), in his important study *Food in early Greece*, summarised the information available half a century ago on prehistoric Greek animal husbandry. Despite the virtual absence of properly published bone reports, he surmised that sheep and goats were numerically dominant over pigs and cattle at nearly all sites, and made some surprisingly modern-sounding remarks about the effects of landscape clearance on the relative importance of different species (1936:65). On the extent and nature of livestock herding, however, he was able only to offer the suggestion that the practices were 'not very different from those found in the same countries today', adding that structures found high in the hills might represent summer herding camps akin to those of contemporary shepherds. (The sole example mentioned by Vickery – the site of Kavousi in Crete – certainly can no longer be interpreted in this way, as subsequent excavation has shown.)

A variety of archaeological evidence discovered since then has been used to infer the existence of specialised pastoral economies in prehistoric Greece. None of it, taken in isolation, suffices to convince the determined sceptic. I shall return below to the direct testimony of excavated faunal assemblages, but it may be noted here that even where the quality of the data is good, the interpretation of it in economic terms is often flawed. The Early Minoan II site of Myrtos (Fournou Korifi) on the south coast of eastern Crete serves to illustrate the point. The complete excavation in the late 1960s of this tiny (0.09 ha.) settlement produced plentiful *in situ* finds of spinning, fulling and weaving equipment, and the animal bones, of which 90% were those of ovicaprids, likewise appeared to indicate wool production (Jarman 1972); the excavator thus felt emboldened to write of the site as 'a textile town, 4500 years ago', which presupposes a specialised animal economy involving substantial herds of sheep (Warren 1968). The final report presented a more sober evaluation of this industry, in keeping with the scale of the settlement (Warren 1972). Later detailed reanalysis in fact, has revealed that the site comprised no more than 5 or 6 non-specialised household units, and that the distribution of textile-related artefacts was more consistent with 'traditional' domestic production at the household level, than with 'industrial' specialisation for exchange via a wide market (Whitelaw 1983).

This same site has been used by Watrous (1977) as an instance of what he would regard as a widespread pattern in Aegean prehistory – that is, a diversified economy involving an element of transhumance and seasonality. Watrous noted the limited extent of cultivable land in the site's vicinity, the absence of water storage facilities, the lack of finds suggesting summer fishing (even though the site is on the coast), and the extremely unpleasant climate there in summer months; this negative evidence, seen in the light of traditional agricultural practice and of patterns of Minoan settlement in the region, indicated to him that Myrtos was occupied primarily in the autumn and winter months, the inhabitants 'moving in the spring to an upland area where they would have grazed their herds and harvested cereals during the summer.' Similar arguments were deployed in proposing that the Dark Age site of Karphi, at 1100 m. above the Lasithi Plain of Crete, was a summer community. Since the site is well above the altitudinal limit for olive growing (600-800 m.), the olive remains found at the site cannot have been grown nearby; conversely, the severity of the winter climate would preclude upland grazing and the survival of autumn lambs. An 'inverse' transhumant pattern was suggested, with a winter descent to land and villages several hundred metres lower in the valleys below the site to the north; Watrous (1977:n.8) argued that a number of Dark Age Cretan sites apparently occur in pairs with this kind of altitudinal separation. In view of its altitude, size, architectural crudity, and paucity of finds, the Early Minoan I-II site at Debla in western Crete (Warren and Tzedakis 1974) could well be one of our best-documented instances of an upland shepherding camp linked, as part of a seasonal cycle, to settlements elsewhere lower down.

This notion of the *seasonal complementarity* of settlements, as a strategy of economic diversification and risk-spreading, has been explored in greater detail by Bintliff (1977: *passim*, esp.116-17). He pointed out, quite rightly, that such systems are not restricted to transhumant herding, but also encompass, for instance, summer trips to distant fishing grounds, or the typical Greek 'Ano-Kato' pairing of villages in two contrasted arable zones or in coastal and inland locations. In practice, however, Bintliff was unable to muster any direct evidence for sheep-goat transhumance in prehistoric Greece. Even so, he expressed surprise that there are those who doubt the antiquity and continuity of the practice, asserting simply that 'such moves with the flocks are essential for the maintenance of large herds in Greece.' This may be so, but it cannot be demonstrated for prehistory merely by citing instances from the Classical sources and in accounts by early travellers and modern ethnographers.

Of more interest, perhaps, is Bintliff's comment (1977:116-17) that transhumance 'can be of great importance in bringing small-scale communities into touch with wider regions, novel customs and artefacts', so that 'a small village may often exhibit paradoxically extensive parallels in assemblage, and even in raw materials, with sites considerable distances away'. If this could be shown to be the case in prehistoric Greece, then it would offer an important indirect means of measuring the extent of mobile animal economies in the past.

That challenge has been taken up recently by Jacobsen (1978; 1984) and Cullen (1985), with reference to the Neolithic period in the northeastern Peloponnese. Cullen's detailed studies of design style on Middle Neolithic painted Urfirnis pottery from sites in this region have shown that stylistic similarity does not always correlate with distance in a straightforward way, and that sites separated by substantial distances (e.g. the Franchthi Cave in the Southern Argolid and Corinth) may produce pottery more similar than sites only a few kilometres apart. She argues that neighbouring villages in competition over resources may well choose to stress *differences* in design style, whereas distant villages engaged in symbiotic relationships may produce deliberately *similar* styles, or act as nodes in an exchange network. Jacobsen, accordingly, has suggested that seasonal pastoralism acted as a unifying mechanism and as a means of cultural exchange with this region of Greece during the Neolithic period. He regards the likelihood of that suggestion as strengthened by the existence today of considerable movement and interaction among three pastoralist groups (Sarakatsani, Arvanites, and Valtetsiotes) who, in different ways, exploit the same area of the Argolid, Corinthia and Arcadia in which these Neolithic sites are located (Koster 1976; 1977; Koster and Koster 1976; Chang 1981). But as Chang and Koster (1986:121) remark, 'actual evidence of pastoralism must be based on positive identification of pastoral architecture or other features.' Nor would transhumance directly *explain* the observed stylistic distribution of pottery, since it is not related to parameters of ceramic production and emulation (Cullen 1985:96).

A rather similar argument has been offered by Bintliff (1977:148-55, 632) on the basis of the class of sites known as Peak Top Sanctuaries in Minoan Crete. These cult locations, which are somewhat similar to the later mountaintop ash-altars to Zeus, are distributed widely throughout the island on, or a little below, mountain peaks in areas of upland pasture; they usually consist of an extensive accumulation of ashy deposits containing burned bone, together with very large numbers of whole or fragmentary votive terracotta models representing humans and animals, especially ovicaprids and cattle. Both their topographic and environmental setting, and the character of the votives found there, have encouraged the orthodox view that they 'came into existence mainly to relieve the fears and cares of the shepherds and cattle breeders' of early Palatial Crete

(Rutkowski 1972:185). A pastoral emphasis to the cult does seem likely, since it is probably from the time of the first emergence of the Minoan palaces (ca. 2000-1900 B.C.) that we should trace the beginnings of a politically-inspired economic demand for the products of large-scale sheep rearing (see below), and there is indeed a clear palatial connection in the nature of cult activities at Peak Top Sanctuaries (Cherry 1986:29-32). Bintliff (1977:630 and 653, map 7) has further suggested that ceremonies held at such regional sanctuaries would serve a territorial and integrative role in embracing large numbers of lowland people who shared an economic interest in the upland pastures for which a particular sanctuary acted as the focus. Once again, the existence of modern or recent transhumant routes, in this case uniting the mountains of Psiloritis, the Messara and the coastal chain of the Asteroussia, is brought into play as confirmatory evidence. Koster (1977) has noted the association of hilltop shrines with pastoralism in Greece today.

Artistic representations of animals, for votive or other purposes, exist in large numbers from prehistoric sites in Greece, and they ought to provide some information about animal husbandry. However, they need not – and, to judge from the surviving animal bones, do not – reflect the relative importance of different species. The prominence of bulls in Minoan cult and iconography, for instance, scarcely indicates a cattle-based economy: in fact, expensive-to-maintain oxen were sufficiently rare that they were named individually on the Knossos Linear B tablets (Chadwick 1976:127). Equally, it would be rash to take the new emphasis on wheel-thrown bovid figures in post-palatial 12th century B.C. votive contexts in Crete and elsewhere as a direct reflection of a shift from mixed farming to a predominantly cattle-breeding economy (Warren 1986:156). A Middle Minoan I bowl from Palaikastro, containing tiny terracotta models of a large flock of sheep with their shepherd (Branigan 1970:fig. 8), is the only unambiguous depiction of animal herding in Greek prehistory; the shepherd is rather better represented in the artistic output of later historical times (e.g. Jost 1975). From the economic point of view, zoomorphic models serve mainly to illustrate the uses to which particular animals were put and to provide a *terminus ante quem* for their first appearance, and they do not cast much light on the structure of the animal economy itself.

Transhumance in prehistoric Greece?

From this brief sketch, it should be apparent that the evidence for large-scale specialised pastoralism in Greek prehistory is both exceedingly meagre and, generally, of an *indirect* or inferential character. *Direct* evidence, in the form of faunal assemblages or excavated pastoralist settlements, is lacking. In the case of the former, this is largely because, until quite recently, archaeozoologists concentrated more on species lists or exotic fauna (e.g. Boessneck and von den Driesch 1981) than on herd structure and composition (Payne 1985). In the latter case, there is a more fundamental problem – the lack of an 'archaeology of pastoralism' (Chang and Koster 1986), i.e. well-grounded theory linking pastoralism as a socioeconomic adaptation to specific aspects of material culture, such as a spatial arrangement distinctive of pastoralist (but not other) sites. Classical Antiquity, too, suffers from a similar paucity of information on the subject: as Georgoudi (1974:158) noted, 'la pauvreté des documents est frappante en tout ce qui concerne les éleveurs et les bergers, leur mode de vie, leurs installations, leurs "mouvements" dans l'espace et le temps.'

Lacking hard data, archaeologists and historians alike have tended to argue on the basis of certain unquestioned propositions about the necessary character of animal husbandry under the ecological and climatic constraints of a Mediterranean setting. The line of argument is one that generations of scholars have unquestioningly accepted (e.g. Braudel 1972:85-102), and it presupposes that the regular movement of men and flocks, whether by some form of 'normal' or 'inverse' or 'mixed' transhumance or by nomadic pastoralism, is an endemic and constant feature of the Mediterranean way of life. In a climate régime of mild wet winters and summer droughts, aridity is the chief check on plant growth in the lowlands, and natural pasture there is usually insufficient for extensive herding in the summer months. The main exceptions are spring-fed irrigated meadows or deltaic marshlands, the latter in particular being the traditional grazing areas for the larger bodied animals (cattle, horses) with substantial water requirements (Wagstaff and Gamble 1982). For the smaller ruminants (i.e. sheep, goats and, less commonly, pigs) to be managed in large numbers, it becomes necessary to exploit upland zones, by carefully timed vertical movements of stock up onto fresh pastures that become available with the melting of winter snows. In Greece, the dissected topography often places these contrasted zones in relative proximity, so that the horizontal movements required are not necessarily very great – certainly less than those typical of pastoral nomads in, for instance, parts of the Near East (Barth 1961). On the other hand, the vertical ascent can traverse rough and steep ground, which nimble, light-bodied ovicaprids can manage well; moreover they are well adapted to the shrub communities which predominate in much of the Mediterranean uplands (true Alpine meadows being virtually absent).

This environmentally-determined view of Mediterranean pastoralism relegates to the wings any cultural factors which might account for the origins, and for the relative scale and importance, of mobile animal economies at various periods in the past. Differences in climate (especially the length and severity of summer drought), altitude, topography and vegetation are regarded as the principle sources of variation, and – assuming a relatively unchanging Holocene climate in the Mediterranean – these would be regional, rather than chronological, in nature. With this set of assumptions, it is legitimate – or so it is claimed – to regard transhumant pastoralism as a 'natural' adaptation to Mediterranean environments and to 'retrodict' its existence in the remote past, even though the evidence for it is merely circumstantial or even absent.

The basis for what may be anachronistic projections backward into prehistory has of course been the well-documented transhumant pastoralism of the modern and medieval periods. For instance, the conspicuous and well-defined transhumant routes in different parts of the Mediterranean – the *cañadas* in Castile, the *camis ramaders* in the Eastern Pyrenees, the *drayes* in Languedoc, the *carraïres* in Provence, the *tratturi* in Italy, the *trazzere* in Sicily, the *drumul oilor* in Rumania – have been studiously mapped (see Braudel 1972:fig. 7, after Müller 1938, for an overall view). These networks, in turn, have been inspected by archaeologists hopeful of finding significant correspondences with distribution patterns of prehistoric sites or artefact types (e.g. Davidson 1980; Walker 1983; Geddes 1983; Barker 1975; 1981; Sterud 1978; Hammond 1976; also Prag 1985, for the Levant). In Greece, there has been a longstanding scholarly interest in the various Balkan nomadic pastoralist groups, notably the Vlachs and Sarakatsani, so that there exists a varied literature on the cultural ecology, history, material culture, language, and folklore of traditional mountain communities in this region (Wiegand 1894; Wace and Thompson 1914; Höeg 1925; Campbell 1964; Kavadias 1965; Hammond 1976; Koster and Koster 1976; Koster 1977; Chang 1981; Winnifrith 1987). Historians interested in Greek animal husbandry in antiquity have been mining this rich vein for some time, but its use as a source of ethnographic analogies by prehistorians is more recent. Most notable are the studies of Eric Higgs and colleagues working in northwest Greece, who used modern pastoralists in the area as a direct model for seasonal transhumant patterns among the Upper Palaeolithic hunters who occupied sites such as Asprochaliko and Kastritsa in Epirus (Higgs 1976; Higgs *et al.* 1967; Higgs and Jarman 1972). Hammond (1976), likewise, has sought to identify numerous migrations and invasions in the northwest Balkan peninsula throughout prehistory, by means of detailed comparison with documented movements of the Vlachs and Saracatsani in the same region during the last millennium. Kirsten (1983), too, attributed the origins of the northwest Greek and Doric dialects to the penetration of sedentary Dark Age communities by transhumant or nomadic shepherds. Höeg (1925) went so far as to argue that specialised pastoralists in Greece today have descended directly from their equivalents in classical and even pre-classical times. These are instances of what Lewthwaite (1981:60) recently termed 'the myth of archaic pastoralism.'

All these writers take it for granted that (1) pastoralism in Greece is largely environmentally determined, and (2) that the environment remains essentially unchanged from antiquity. Both assumptions have lately been called into question, not only with respect to prehistoric Greece (Fotiades 1980; Halstead

1984; 1985; 1987), but also more widely (e.g. Garnsey 1988; Lewthwaite 1981; 1982; 1984a; 1984b; Walker 1983).

Taking the second point first, discussion has centred on a problem in historical ecology: were the upland pastures, which are vital for a flourishing transhumant economy, actually available on the same scale in the past? At present, data from palynological and ecological studies necessary to answer this question with certainty simply do not exist in sufficient abundance. However, it has been clear for some time (e.g. Turrill 1929; cf. Wertime 1983) that the summer pastures that now exist in many highland areas, not only in Greece but more widely throughout the Mediterranean, were created (or, at least, have been greatly expanded) through the human interference of grazing and lumbering activities in quite recent times. Historical evidence for *deliberate* deforestation to extend pastures for large-scale herding in the medieval and post-medieval periods is actually much better in Italy and Spain than in Greece; but other sources of information point in the same direction. For instance, Quezel (1967) has noted the relative paucity of endemic species on the high mountain grasslands of Thessaly, and Halstead (1984:2.7.1) suggests such a paucity may indicate that they are largely a recent development. Well-dated Greek pollen diagrams are still rather few and of variable quality, but they too support the view that most parts of the country in antiquity were far less treeless than today, and that *extensive* deforestation in some mountainous areas, at least, is a phenomenon of the last few centuries (Bottema 1974; Turner and Greig 1975; Turner 1978; Halstead 1984:2.7.2; 1987). Indeed, for the lowland areas favoured by the first farmers in Greece, it is difficult to find clear-cut indications in the palynological record of woodland clearance by early agriculturalists: this might suggest that the process itself was slow and that running large herds of animals in such a well-wooded environment would have been impracticable (Halstead 1981; 1987).

Of course, one must beware of generalising too widely from such admittedly patchy data. The rate of forest clearance and the extent of the spread of upland pasture have probably been locally variable. Although the Greek mountains are not so high and cold as to prevent tree growth, altitude is a significant factor affecting the availability of mountain grassland, as also are aridity, parent geology, slope and erosion. It is quite likely, moreover, that the use of mountain pastures developed *pari passu* with lowland clearance (since the latter would eventually generate a substantial complementary *winter* grazing resource). Even if Halstead (1984:2.7.1) is only partly correct in asserting that in prehistory 'these high mountain pastures would have been much reduced in extent and restricted to the driest limestone plateaux where the modern grass cover is relatively thin' and that 'the richest (and often most accessible) pastures of today, on impermeable rocks, would probably not have existed', the essential point remains – that it is wholly unwarranted to assume a continuous transhumant pastoral tradition solely on the grounds that the environment itself makes this 'inevitable'. Future investigation may throw more light on Halstead's interesting speculation (1987) that the appearance of the distinctively-adapted traditional pastoralist groups in the high Pindos a millennium or so ago (Hammond 1976) should be regarded as the colonisation of a new ecological niche becoming available only at that time.

With regard to the second assumption noted earlier – that mobile economies are a categorically efficient adaptation to areas of complementary resources – a thoroughgoing critique is to be found in a series of papers by Lewthwaite (1981; 1982; 1984a; 1984b), and a number of other recent authors on the archaeology of pastoralism clearly feel some sympathy for his viewpoint. He rightly points out (1981:57) that many archaeologists have expected *as a matter of course* to find evidence of transhumant economies in the data, without asking the more fundamental question of why large-scale ovicaprid herding should take place at all. After all, pigs convert herbaceous growth into protein more efficiently (as the West Mediterranean acorn-hog economies demonstrate); dairy pastoralism, at least as an activity supporting whole communities, is so labour-intensive and space-extensive as to be unlikely, except where articulated with large market economies; wool took on greater significance only during and after the Roman-Early Medieval periods, when fine-woolled breeds appeared (Ryder 1983). Furthermore, the adoption of a seasonal lowland-upland system incurs significant energetic costs, since both transhumant movements and the grazing of poor pasture involve work for the animals. 'Complementary' mountain grasslands may be convenient, but there are inherent disadvantages in a system dependent on pasture yields at high altitude: for instance, Baticle (1974:84) reports that yield levels on the plain are reduced fivefold at 2000 m.

Thus, while *localised* stock movements between ecologically contrasted zones would make obvious sense in a wide variety of situations, long-distance transhumance is in many respects awkward and inefficient, and it is pertinent to examine the economic and sociopolitical contexts in which it has flourished historically and which have provided the most influential controlling models for archaeologists. The classic instances are those of post-medieval Spain and Italy – the Mesta of Castile (Klein 1920) and the Dogana della Mena delle Pecore in Puglia (Sprengel 1971) – both of which, at their height in the 16th-17th century, involved the management of some 4-5 million sheep. State-sponsored and state-controlled, they depended for their existence on the European-wide wool trade which provided a massive and relatively constant demand unlike anything previously. To run sheep at this scale was feasible given the low labour costs of a management strategy devoted solely to wool (sheep to shepherd ratios of ca. 1000 to 1: Baticle 1974), and an important prerequisite was the existence of a rich élite owning or controlling huge expanses of grazing land. That political circumstances strongly affect the very possibility of horizontal transhumance is equally apparent in the Roman world, though of course there is an aristocratic bias in the sources that inform us about it (Skydsgaard 1974; Garnsey 1988). In either case, it is access to land and to markets for pastoral products, both on a large scale, that seems critical; neither circumstance is likely to have existed in the prehistoric world, and the dissolution, in such extreme form, of the links between the arable and pastoral sectors of the economy seems very improbable. In pre-modern and modern times, too, it is quite clear that the degree of pastoralisation of different parts of the Mediterranean is largely a function of period-specific political and economic factors, such as the size and stability of world market demand for specific types of ovicaprine cheeses, or the availability of land through the decline of traditional agriculture, or fluctuations in population size and distribution, or the relative strength of political authority (cf. Davis 1977:20-28; Fotiades 1980).

Lewthwaite (1981:61) may be overstating the case in proposing 'that transhumance be regarded as a tactical adaptation deployed in the peculiarly difficult circumstances of the medieval Mediterranean, rather than a long term strategy'; Chang and Koster (1986:102-4) certainly think so. Nonetheless, his main point seems soundly based: there has been a good deal of inflation in estimates of the importance of large-scale transhumant pastoralism, both for classical antiquity and for prehistory, and this has served only to obscure the importance of small-scale transhumance in 'traditional' Mediterranean farming systems. Mesta- or Dogana-type economies were not a 'natural' response to the environment by ruggedly independent shepherds, but highly specialised, institutionalised systems of economic control organised by, and for the benefit of, an exploitative élite. To project them back into other periods with far more restricted exchange or marketing arrangements and with less hierarchical political systems is wildly anachronistic. In prehistoric and protohistoric Greece, the only period at which such an analogy might be relevant is that of the developed Minoan and Mycenaean states: appropriately enough, it is precisely in the Mycenaean Linear B tablets of the 13th century B.C. that we find our only textual indications of specialised pastoralism on a massive scale and with a strong element of state intervention (Killen 1964; 1985). In general, however, the large-scale pastoralist systems of medieval and modern times developed in response to the demands of a 'world economy' for wool or dairy products and involved shepherds in a symbiotic dependency on urban markets unlike anything known in the prehistoric Aegean.

The archaeology of pastoralism and the relevance of ethnoarchaeology

It thus seems clear that much of what has been written about the place of pastoralism in Greek prehistory has surprisingly little empirical or logical foundation. The fact that the demographic, economic, and socio-political conditions which have favoured transhumance in recent times cannot be shown to have operated in prehistory does not, of course, provide grounds for excluding the existence of pastoralism altogether. Quite simply, at present we have no means of assessing its nature and relative importance.

What is most striking is the absence of *direct* archaeological evidence. Who has actually excavated a prehistoric site in Greece that is generally agreed to be a pastoralist settlement or a transhumant transit site? Are any published faunal assemblages unambiguously identifiable as the outcome of specialised pastoralism? Most statements on the topic in fact involve inferences at several removes from the hard data. For example, sites at high altitude, or seasonally occupied, or containing specialised artefacts (such as animal figurines or statuettes of relevant tutelary deities), or in proximity to transhumant routes used in modern times, have often been supposed to imply pastoralism; but none of these features automatically *demands* that interpretation, and the arguments linking archaeological observations to such assigned meanings remain tacit. This is just as true for prehistoric and Roman Italy as for Greece: regional site patterns, hilltop shrines, and the distribution of allegedly pastoral artefact types feature prominently in the work of Barker (1975; 1981) or Pasquinucci (1979). (Oddly, the literature for certain parts of the West Mediterranean, such as Sardinia, seems to suggest the existence of much better on-the-ground indications of prehistoric pastoralism – excavations of seasonal/transit caves and rock shelters in upland zones, for instance, or the widespread occurrence of artefacts connected with dairy processing or textile manufacture (e.g. Lilliu 1975); but Lewthwaite (1984a:253-6) has shown that this is largely a bias in sampling and interpretation, stemming from a fixation with the nomadic *pastore-guerriere* among Sardinian prehistorians.)

Should one conclude, then, that prehistoric pastoralism is simply inaccessible archaeologically? For Greece, at least, such a conclusion would be premature. As Chang and Koster (1986:111-12) point out, pastoralists depend on dispersed and seasonally available resources, some of them in zones marginal or wholly unsuitable for agriculture; the functional specialisation and brief occupation-span of many pastoral sites may result in relatively ephemeral archaeological traces. These factors demand the use of a regional site-location strategy, embracing upland areas well outside the lowland plains and main population centres which have been the traditional focus of archaeology in Greece, as well as intensive search tactics. Archaeological surveys satisfying these requirements are still quite rare in Greece (Cherry 1983), so that the absence of prehistoric sites with clear indications of a pastoral economy has no necessary significance.

There is a more fundamental difficulty, however. This is the lack of unambiguous criteria for distinguishing the material remains of a pastoral society and economy from an agricultural one (or, indeed, one based on foraging and hunting). What sorts of sites and material culture would be the result of pastoral activities? Large quantities of faunal remains, or the age-sex structure of the bones, or specific artefacts, all may indicate heavy engagement in animal husbandry, but they do not establish the economic identity of the people involved; the literature (particularly that for Dark Age Greece) is replete with singularly weak arguments that particular house-forms, or burial rites, or crude pottery types are 'characteristically pastoralist'. It has become increasingly apparent that the relationship between pastoralism as a socioeconomic adaptation and the material culture resulting from it is one best investigated in a contemporary setting – that is, by archaeological observation of living pastoral groups in action. A good example is afforded by Murray and Chang's (1981) study of a complex of modern animal folds and pens within a natural limestone sinkhole west of the village of Didyma in the southern Argolid of Greece. They were able to compare their inferences from detailed archaeological description of structures and artefacts at the site with ethnographic observations of the site in use and with information supplied by the herders themselves. The relationship between pastoral behaviour and its material correlates, the processes of site-formation, and the limits to inference are all particularly clear in this case-study.

Such *ethnoarchaeological* research is now becoming more widespread in a number of areas, especially in parts of Africa, the Near East, and around the northern shores of the Mediterranean (e.g. Aurenche 1984; Bray 1983; Cribb 1982; 1983; Gifford 1978; Hole 1978; 1979; Solecki 1979; for a general review, see Chang and Koster 1986, especially 127-30). In Greece, the earlier folkloristic and linguistic, or purely social anthropological, studies of transhumant pastoralists have begun to be supplemented by more materially and ecologically oriented research, undertaken either by archaeologists or by cultural anthropologists operating under the umbrella of interdisciplinary, regional archaeological projects (e.g. Chang 1981; 1984; Koster 1976; 1977; Koster and Koster 1976; Murray and Chang 1981; Chang and Koster 1986). Similar data, though not collected by archaeologists, are also available for some parts of the west Mediterranean (Angioni 1974; Cristofini *et al.* 1979; Lewthwaite 1984b).

Considerable care is necessary in using ethnoarchaeological work in a way that is not merely the uncontrolled imposition of present circumstances onto the prehistoric past (e.g. Jacobsen 1984). Nonetheless, it is clear that there is a wealth of potential information on a wide variety of different types of pastoralist site and on the facilities and artefacts used at them – not only structures for housing shepherds and for sheltering or confining animals, but provision for water and for storage of fodder or animal products, as well as ritual or burial sites, and even improvements to animal trails and pasturelands (Chang and Koster 1986:111-13). Because pastoralists often employ organic, perishable objects, or take them with them on leaving a temporarily-occupied site, or make expedient use of locally available materials, excavated artefact inventories at putative pastoralist encampments are notoriously difficult to interpret (e.g. Robbins 1973; Hole 1974). Ethnoarchaeological studies of patterns of use, manufacture, and discard of artefacts by herders are thus of considerable importance. Likewise, modern studies offer valuable insight on the environmental and ecological factors constraining the location of pastoral sites, and on the spatial arrangements of artefacts, structures, and activity areas at various categories of site. What, for instance, distinguishes winter from summer/spring camps, and both from transhumant transit sites (Hole 1978; 1979)? How widely is it the case that sheep and goat herding can be discriminated on the basis of variables such as fold sizes (Chang 1984:fig.3) or the shapes of milking enclosures (Lewthwaite 1984b:31)? Above all, what *archaeologically preservable* aspects of pastoralist economies are sensitive indices of the degree of specialisation and intensification? Of particular value for this latter question are recent attempts to use simulation and computer modelling of modern or idealised herd structures, under different types of exploitation strategy and kill-off patterns, as a means of interpreting prehistoric mortality profiles (Dahl and Hjort 1976; Redding 1981; Cribb 1985).

It seems likely that only in this way – i.e. studies providing the links between dynamic behaviour in the past and its static material outcome in the present – will progress be made in studying pastoralism in prehistoric and early historic Greece. The irony is that the raw material for such studies, the traditional pastoralist communities of the Mediterranean, is dwindling rapidly at the very time that ethnoarchaeological and anthropological interest in it is on the increase.

Long-term changes in animal husbandry in prehistoric Greek economies

So far, I have been negative about the evidential basis for pastoralism in prehistoric Greece. At the same time, however, it seems that the new breed of intensive regional surveys, the development of ethnoarchaeological studies of pastoralist societies, the growing sophistication of interpretative methods in archaeozoology, and a general growth of interest in the 'archaeology of pastoralism', all provide grounds for considerable optimism about future developments. Meanwhile, where do we stand? Can anything be said about the changing rôle of animal husbandry in the subsistence economy as a whole and, specifically, about the wider context in which specialised pastoralism may have become a differentiated sector in later prehistory? To tackle this question, I propose in the remainder of the paper to contrast three very different sets of circumstances: the first farmers of the earlier Neolithic, the large-scale herding in the Mycenaean world, and the communities of the Dark Ages. A connected 'narrative account' of animal husbandry during this very long period is *not* attempted here, for the simple reason that evidence from the Early and Middle Bronze Age is still of very poor quality. This is unfortunate, for it is precisely in the period of state emergence that the differentiation of a pastoralist sector is most likely, if the Near Eastern sequence offers any guidance (Lees and Bates 1974; Briant 1982).

Faunal and botanical samples from early farming sites in Greece are still quite rare (Payne 1985; Barker 1985:64, Table 6). On the other hand, enough is known of environmental conditions, and of the spatial distribution and typical sizes of the settlements of the first farmers, to be able to build realistic models of subsistence systems at that time. This has been done most convincingly by Halstead (1981; 1984). In brief, his argument is that 'in Greece, the concentration of Early Neolithic population in the well-wooded lowlands makes an animal-based economy quite unlikely for most of these early groups' (1981:314). Hypothetically, to satisfy the calorific requirements of a Neolithic *tell* community of between 40 and 240 people wholly or largely with ewe's milk or mutton, for instance, would require sheep herds so large as to be impossible to run in such a densely forested environment, and whatever the availability of open upland vegetation, its exploitation would have been hampered by very limited complementary lowland winter grazing areas. Plant foods, mainly cereals and pulses, were the dietary staples, and there exist several convergent lines of argument to show that cultivation was both highly productive and very localised around the settlement (perhaps 10-50 ha.). Such intensive crop cultivation, probably with simple rotation systems, was possible partly because proximity to the settlement allowed fertilisation of the fields with domestic refuse and human manure, but also because fallow weeds and crop stubble would be an essential grazing resource for animals kept in what amounted to small clearings in the deciduous forest, and their manure would be an important contribution to the soil nutrients necessary to maintain high yield levels. A specialisation in husbandry involving large numbers of animals would require either the provision of leafy fodder in huge quantities, or the physical removal of the animals to grazing areas at some distance from the settlement – and thus the loss of their manure. It is often not appreciated that the sheep is a phenomenal producer of manure (about 10 times it own weight per annum), and Lewthwaite (1981:61) has gone so far as to argue that, in areas of cereal agriculture on poor soils, dung was the principal reason for keeping ovicaprine flocks. Certainly, sheep and goat always make up at least two thirds of the animal bones at early farming sites in Greece (Bökönyi 1973; Jarman and Jarman 1968), and their age/sex structure indicates meat, rather than wool or dairy products, as the main product (Halstead 1981; 1984; 1987).

Chang and Koster's (1986:102, 108) recent criticisms of this interpretation do not seem well-founded. Most flocks were no doubt kept for a variety of purposes, as they claim, but they do not seem to have grasped the severity of environmental restrictions on the possibilities for animal husbandry in the 6th and 5th millennium B.C. Halstead's model, in which modest flocks of ovicaprids clean and fertilise the cultivated fields near the settlement and – in the process – convert unusable plant materials into meat and milk serving as dietary supplements or fall-backs in times of crop failure, harmonises well with the basic pattern of use of the 'package' of domesticates in the early stages of their introduction in the Near

East (Flannery 1969). It may be noted that, where grazing is limited, Payne's (1973) milk, rather than meat, strategy would yield most energy, so the fact that this does not seem to be the actual strategy followed perhaps indicates a concern more for risk minimisation than the maximisation of energy capture. 'The herding strategy characteristic of neolithic Thessaly, therefore, offered a compromise between energy productivity and security, while the improvement in male survivorship in the Bronze Age suggests increasing emphasis on wool and/or herd security at the expense of energy productivity' (Halstead 1987).

It seems likely, then, that early farming in Greece involved essentially the same components and strategies as did peasant subsistence farming in the classical period: a mixed-farming system, with cereals and pulses as staple crops, and a modest number of the smaller ruminants kept as a multi-purpose supplementary resource for use at the household level, and as an emergency food 'bank.' There is every reason to regard this kind of animal exploitation – small-scale, unspecialised and underdeveloped, symbiotic with the arable sector, and mixed in character – as a virtually constant feature in most parts of the Mediterranean ever since the introduction of domesticates. But in later prehistory and historical times, there is evidence of a far wider range of uses of animals, some of them involving semi-autonomous, specialised pastoralism on a spatial scale sufficient to require formal arrangements for access to grazing rights in the territories of other polities. We need to ask, then, by what processes and at what stage the transition from Neolithic mixed farming/herding economies to pastoral specialisation occurred, and whether it too became an indigenous and ubiquitous feature of the Greek landscape and economy.

This is a large question, since most authors are agreed that pastoralism *as a specialised economic form* has developed historically in the context of the formation of state societies (Khazanov 1984), which is itself a process not well understood. The best discussions so far (e.g. Lees and Bates 1974) have been with reference to Mesopotamia. Nonetheless, relevant data from Greece are not lacking.

One important piece of evidence is the shifting balance of stock ratios throughout the five millennia of the Greek Neolithic and Bronze Age, as reflected in surviving faunal assemblages. Although the quality of the information is patchy and coarse, and thus not immune from criticism (e.g. Payne 1985), there is a clear trend in the early and middle Neolithic – at least in Thessaly, Macedonia and Crete, where the best data exist at present – for a general increase in the numbers of cattle, or pigs (or both), at the expense of ovicaprids (Halstead 1981:fig. 11.2; 1984; Barker 1985), a shift in emphasis that corresponds with a diversification of crop systems also. There is little room for doubt that this trend is a reflection of man-induced changes to the landscape itself: woodland clearance for leafy fodder, firewood, or the expansion of arable fields, accompanied by the development of secondary browse in place of mature deciduous forest, and by decreased woodland regeneration, or in places even by soil erosion. Such changes in the relative proportions of the types of browsing/grazing most suited to the different domesticated species, together with changes in their proximity to settlements, would alter the cost/benefit ratios of different permutations of stock.

It was also in the period c. 5000-3000 B.C. that population growth led to the expansion of Neolithic and Early Bronze Age settlement, as a process both of infilling existing areas of established agriculture on the fertile, easily-cultivable soils, and of colonisation of new zones. The latter included environments such as the drier parts of the Thessalian plain (Halstead 1977; 1984), the Aegean islands (Cherry 1981), and more rugged and less fertile areas in southern Greece and Crete. In these more marginal areas, crop production would be considerably less predictable and reliable, so that animals probably took on increased importance as a hedge against uncertainty and a wider range of strategies for exploiting them was followed – or so it would seem from the somewhat limited faunal data presently available for the later Neolithic of Greece.

But other factors were at work too. In particular, the progress of deforestation, and the creation of a more open landscape with greater amounts of land devoted to arable agriculture, very likely allowed more livestock to be kept and encouraged their segregation for at least part of the time. Indeed, it is probably to this period that we should date the origins of husbandry systems, at least in some parts of Greece, in which animals made a substantial dietary contribution. It is also the time at which the first (admittedly weak) signs of *local* transhumant pastoralism perhaps appear. One instance may be the use of summer grazing on the lush Peneios river floodplain in Thessaly or on the Drama plain in Macedonia (both in later times traditional areas of cattle raising), a development possible only with the creation of more extensive, complementary winter grazing. Another is the widespread occupation during the late Neolithic of caves in parts of southern Greece more suited to herding than agriculture; but it is very uncertain from the archaeological evidence whether these were perennially occupied by small groups of dairy pastoralists or by seasonally transhuming members of lowland farming settlements (Halstead 1981:326-6). This is a good example of the type of problem where intensive surveys and ethnoarchaeology can be expected to shed light.

Whether or not these suggestions are correct, it seems virtually certain on various grounds that further specialisation in stock-keeping would be favoured by the widespread trend, from the late Neolithic and throughout the Bronze Age, towards larger, nucleated settlements. Gamble (1982) has explored some of the implications of nucleation on the costs and organisation of animal husbandry. Longer journey-to-field times and larger fields encouraged the spatial segregation of crops with different labour demands (as increasingly pure crop residues possibly suggest), and provided the incentive for the keeping of load-bearing animals such as the donkey and horse (osteological evidence for which first appears in the Greek Early Bronze Age). The use of oxen as traction animals must have been as limited for subsistence farmers in prehistory as it was in classical and Roman times, given the expense of maintaining them and the difficulty in some environments of providing for their prodigious water requirements; but the need to feed larger populations including specialists of various kinds, albeit by less efficient and productive methods of agriculture, probably gave importance to ox-power in some areas. For the same reason, the taking in of more land under cultivation must have generated an imbalance of winter/summer grazing on the fallow and stubble, a problem soluble only by the physical removal of animals to summer pastures located beyond the immediate confines of a settlement's farmed land. Once this circumstance existed, together with the opportunites for economic specialisation and exchange within and between communities in the later fourth and third millenia, there would be every reason to keep flocks of the more mobile ruminants (sheep/goats) in substantial numbers and to manipulate them so as to maximise wool, rather than milk or meat, production. That, certainly, is what the bones at sites of this period from Crete to Macedonia do suggest: a marked increase in the proportion of sheep and goats, larger numbers of animals kept into adulthood for breeding, bigger sheep, and more male sheep (which produce more wool per unit of bodyweight), all indicate that wool production was a priority.

All these developments have to be seen in the wider context of similar trends and changes in animal husbandry throughout Europe and the Near East in the period very roughly four to five millennia ago – what Sherratt (1981; 1983) has very appropriately termed the 'secondary products revolution.' By this he means a range of uses and products of domesticated animals which were not part of the original complex of plant and animal domestication developed in the Near East (and typifying the intensive hoe-gardening of earlier Neolithic farming in Greece). They include ox traction for ploughing and equids for pulling carts; the secondary products of ovicaprids, especially wool and other textile fibres; and the continuous production of milk and various storable milk products. Some of the economic intensifications brought about by this 'revolution' have already been noted. Ploughing with animals increased production and, by making agriculture viable on poorer soils, promoted colonisation of hitherto marginal areas; the ox-cart, the pack-donkey and the horse facilitated exchange and bulk transport, and made new types of site location economic; wool and other animal-fibre textiles, together with increased use of milk, provided exchangeable products from non-agricultural or marginal land, and encouraged 'the development of the pastoral sector with transhumance or even nomadism' (1981:262). Thus, if one wished to isolate the 'origins' of pastoralism in Greece, it would be in this important transition between 'an initial stage of hoe cultivation, whose technology and transportation systems were based upon human muscle-power, and in which animals were kept purely for meat...and a second stage in which both plough agriculture and pastoralism can be recognised, with a technology using animal sources of energy' (1981:263).

The social implications of these changes beyond subsistence and self-sufficiency have not been lost on a variety of recent authors (e.g. Renfrew 1972; Gilman 1981; Cherry 1984). It seems likely that it was the increased costs of animal husbandry contingent upon nucleation of settlement that promoted the running of much larger, consolidated flocks, in addition to (or perhaps at the expense of) small-scale, local herding by individuals (Gamble 1982; Halstead 1987). The Roman and Medieval parallels show that it is frequently the élites who finance, control, and take the profits from such enterprises, and the recurrent conflicts between peasant farmers and pastoralists assume a social dimension for that reason. Was the same true of prehistoric Greece? In general, the main trend that can be discerned is the evolution of overall farming strategies to reduce the risk of subsistence crisis in an inherently risky environment. Animals, however, offer a very convenient means of 'banking' the normal surplus in such systems, to form an important contingency reserve which may not in fact need to be called on. Where that was so, opportunities existed for stockpiling wealth in herd form. And with the increasing emphasis on specialised husbandry, the way was open not only for the accumulation of individual capital wealth, but also for capital intensification of the sort seen most clearly in the Minoan and Mycenaean palace societies (Cherry 1984; Gamble 1982). Paradoxically, the removal of many members of these societies from food production (again, largely as a function of the nucleation of population), and the apparent trend towards an agrarian régime based on cereal monocropping, increased the significance of livestock as a back-up food reserve – a situation in which we might well expect the development of an élite monopoly. The idea that Mycenaean state interests in sheep-raising and wool reflect monopolistic banking, arising in the first instance out of the need for security against crop failures, has been explored in most detail by Halstead and O'Shea (1982): in their view, state power resided in the ability to guard and manipulate stored foods, whether kept in palace magazines or on the hoof.

In Greek prehistory, therefore, it seems likely that it was only under the unusual and relatively short-lived circumstances of the palatial economies that there emerged large-scale, specialised pastoralism of a kind akin to the Mesta and Dogana, or the systems controlled by certain Roman aristocrats. The Linear B archives, both at Knossos in Crete and at Pylos on the Greek mainland, provide evidence, unrivalled in detail, about herding activities on a massive scale, the products of some of which were transformed, via further inputs of (tied or slave?) labour in palace workshops, into high-value/low-bulk textile commodities which to some extent penetrated the élite exchange networks of the eastern Mediterranean. The details of the sheep and wool tablets, and the unravelling of the management strategies that they cryptically document, are well known and need not be repeated here (Ventris and Chadwick 1973; Killen 1964; 1985).

What is, as yet, less well understood is the degree of palatial control over this highly specialised pastoralist sector, and its rôle in the economy as a whole. The Knossos tablets record production quotas against the names of places and/or persons for flocks spread widely throughout central and western Crete, and totalling at least 100,000 head. The precise relationship of shepherds, or 'owners', or 'overseers', to the palace is far from clear; but Bennet (1986), using new toponymic evidence, has shown that there is a clear diminution of direct palatial involvement as a function of increasing distance from Knossos. Equally, Halstead (1981:332) estimates that the wool from Knossian flocks might clothe 20-30,000 people, whereas the likely population of central Crete at this time – though speculative and much in dispute – was surely considerably larger than that (e.g. Renfrew 1972:255). Thus many sheep may have been outside the interests and control of the palace bureaucracy (or at least receive no record in its archives).

Chang and Koster (1986:110) explicitly assume that this state-organised pastoralism was in *deliberate* conflict with the subsistence autonomy of individual households: 'By offering their flocks the maximum opportunity for expansion by failing to cull males, the élite were not only increasing wool production but also effectively pre-empting grazing resources which could otherwise easily be poached by the flocks of subordinate households. This policy would have presaged attempts by Spanish and Italian élites to effectively appropriate resources and convert them to wealth through organisations like the Mesta and Dogana. Thus, the large palace flocks could have served as storage for wealth, produced a valuable commodity in wool, and helped the élite control marginal resources which in turn could lead to a relatively cheap form of control over the subordinate population.' It certainly seems fair to assume a dichotomy between extensive and large-scale state pastoralism, on the one hand, and subsistence herding elsewhere within the palace territory, on the other.

The Linear B tablets are the only case in the whole of later prehistoric Europe where we have available the details of the organisation of a complex agri-pastoral system. Seen in the light of long-term stability and change in Greek animal husbandry, the specialised pastoralism they document so well stands out clearly as moulded far more by historical circumstance than by environmental necessity. Its emergence required demographic change, the rise of political hierarchy and control, and the existence of outlets for its products. Although the evidence is largely lost in the mists of the Dark Age, comparable instances suggest that its decline would be rapid, once these prerequisites disappeared with the end of the Mycenaean polities.

The problem of the Dark Age

The rôle of animals in the Early Iron Age of Greece is a problem whose difficulties are compounded in two ways. Firstly, several generations of scholars have assumed *a priori* that this era was one characterised by some form of pastoralism – usually unspecified, but involving reduced levels of sedentary arable agriculture and an increased dependence on mobile animal economies. At the core of many such arguments, of course, is the belief that the end of the Mycenaean age witnessed widespread invasions of Dorians and other alleged immigrants (Hammond 1976), and that they enjoyed a 'lower' level of culture than their predecessors in Greece, made necessary in part by their nomadic way of life and explaining the difficulties of their archaeological recognition; direct analogies are often made with the Slav invasions or with the movements of the Vlachs in northwestern Greece. It is not usually made clear that there is an implicit (and demonstrably false) assumption behind such assertions: that sedentism is the desired condition and that a 'reversion' to mobility will occur only when sociopolitical or economic conditions leave no other alternative. This is the romantic image of the pure nomadic shepherd as fiercely independent, stoically resigned, and lying outside the constraints and benefits of lowland, agriculturally-based civilisation – a point to which I return below.

The second problem is simply one of lack of evidence. So far, the *only* excavated Early Iron Age site to have produced a decent sample of animal bones is Nichoria in Messenia (Sloan and Duncan 1978). Other sites, such as Vitsa or Karphi are less satisfactory. Inevitably, the faunal assemblage at Nichoria has come to bear more weight of interpretation than it can stand, and there has been the temptation in some quarters to allow the results from this single site to serve as a proxy measure for the animal economics of Dark Age Greece as a whole. What Duncan and Sloan observed was a sharp rise from the preceding Late Bronze Age levels at the site in the proportion of cattle bones, and a noticeable shift – noticeable perhaps only because the sample size is so pitifully small – in the mortality profiles of both ovicaprids and cattle. This they took to reflect a major reversion to a stock-rearing economy. Unfortunately, without the full integration of all the economic data from the site, and in the absence of comparanda, this is a difficult proposition to evaluate. But in the light of comments earlier in this paper, it is perhaps more likely that these data indicate the abandonment of a specialised ovicaprid wool/milk strategy, in favour of a mixed strategy in which animal meat played the same rôle as a contingency resource, as was the case before the development of extensive, state-controlled pastoralism (Nichoria was probably one of the more important settlements in the Mycenaean kingdom of Pylos, parts of whose livestock records survive).

This general lack of information has, quite naturally, encouraged the pursuit of a wide variety of other types of argument in support of the notion of Dark Age pastoralism. It is surprising how influential the conventional exploits of the meat-devouring, ox-slaying heroes of Homer have been, despite the clear implications elsewhere in the poems that bread is the real staff of life. Given the near-insoluble problems of (1) the date of composition of the Homeric epics, (2) the distinction of anachronistic *genre* 'distancing devices', adding verisimilitude, from real-life contemporary description, and (3) the representativeness and cultural bias of the audience for which the poems were composed, or first written down, it is very difficult to see how Homer can act as a witness for or against the proposition. Likewise, too much has been made of Hesiod's *Works and Days* as 'a tract designed to inculcate the first principles of arable farming for a readership which was long inured to pastoralism' (Snodgrass 1987; cf. 1980). This argument, drawing on Thalia Howe's earlier analysis of Hesiod and presented by Snodgrass in his book *The Dark Age of Greece* (1971:378-80), was originally felt to gain support from the supposed facts that the Mycenaean economy was pastorally-based, and that the earlier Iron Age in Greece was culturally impoverished in a manner characteristic of all pastoral societies. Both propositions now appear factually inaccurate, as Snodgrass (1987) now concedes.

In his book *Archaic Greece* (1980), and more recently in one of his Sather lectures (1987), Snodgrass has continued to urge the proposition that the centuries of the early Iron Age witnessed a marked but temporary increase in the proportion of the Greek landscape that was given over to a pastoral economy. He candidly admits that this remains a hypothesis for testing and usefully sets out the variety of arguments that have been advanced in its favour by scholars of widely divergent persuasions. These therefore do not need to be reviewed in detail here. They include the evidence of patterns of place-name survival and of site continuity or abandonment through the centuries of the Mycenaean and Dark Ages; the frequency of animal-figurine dedications in the early Iron Age levels at a number of Greek sanctuaries, both in the Peloponnese and in north-central Greece; the correspondence between the funerary rites practised by pastoralist tribes in Central Asia and the early form of the Olympic Games; the alleged association of the typically Dark Age apsidal house plan with nomadic pastoralism in other contexts; the increase in the Early Iron Age of hand-made pottery, which some feel to be the appropriate accoûtrement of mobile pastoralists lacking access to the potter's wheel and not producing for large-scale urban markets; and the prevalence of cattle pastoralism and seasonal transhumance during later periods in certain parts of Greece with strong traditions in Dark Age legendary history. Individually, none of these arguments is decisive; nor should the lack of any better evidence seduce us into supposing that *collectively* they support a reconstruction any more secure.

I hope that my own attitude to such suggestions is obvious from the perspective outlined earlier in this paper. *They all founder on the general lack of well-grounded theory linking pastoralism as a socioeconomic adaptation to specific aspects of material culture*, and the problems are only exacerbated by the notoriously thin archaeological indications of that material culture in the post-Mycenaean era. The input of theory-building, based on ethnoarchaeological data, is nowhere more urgent than here. Certainly, an alternative proposition requires consideration. This is that, with the collapse of the Mycenaean polities, the 11th to 9th centuries in Greece witnessed not a wholesale shift to nomadic pastoralism, but a reversion to more localised, intensive, mixed, non-specialised farming systems in which animals served much the same functions as they had before the rise of stratified states in the Aegean – i.e. an essentially agrarian household régime, with varying mixes of animals in small numbers kept primarily as a back-up resource. No doubt there was much regional variation, and animals were kept for quite different purposes and following different management strategies in different areas. Yet the overall conclusion would be that the highly specialised ovicaprid pastoralism (and perhaps the élite monopoly of expensive-to-maintain large-bodied animals, such as the chariot horse or the ox?) which typifies the palace societies of the later Bronze Age was an abnormal adaptation, simply reflecting the short-lived political circumstances of the Minoan and Mycenaean age.

It remains to ask why pastoral nomadism should so widely have been assumed to be the dominant economic mode in Dark Age Greece, when the empirical evidence to make any judgement is so notably lacunose. Much of the answer has recently been supplied in Brent Shaw's illuminating paper '"Eaters of flesh, drinkers of milk": the ancient Mediterranean ideology of the pastoral nomad' (1982-3). It is well-known that nomads, ancient or modern, have never had a 'good press' and that, for classical antiquity, we depend largely on unsympathetic accounts, rarely based on personal autopsy, by historians, geographers, military men and administrators, to whom pastoralists were either an irrelevant curiosity or an irritating impediment. Shaw, however, has shown that these accounts go beyond mere prejudices to enshrine an organised and structurally consistent *ideology* of nomadism, one that has 'continued to function as a part of European thought concerning outsiders to the end of the early modern period' (1982-3:30). Briefly, this ideology revolves around the complete taxonomic separation and contrast of nomadic pastoralists (*nomades*), seen as the ultimate barbaric human type, and 'civilised' sedentary agriculturalists (*aroteres*): economic régimes thus serve as an explanatory model for stereotyped images of human customs and behaviour. From Homer to Ammianus Marcellinus, the pastoralist is defined simply via logical opposition to the essential criteria of civilisation: mobile and without established homes, non-urban, *polis*-less, without properly constituted rules or law-codes, lazy and parasitic (because he does not *work* the land and *harvest* crops, like the farmer), an eater of meat (often *raw* flesh, and even raw *human* flesh) rather than grain, and a drinker of milk, not wine. By a confused social syllogism, nomadic pastoralism comes to bear the full stigma of uncivilised barbarity.

The legacy of this consistently hostile tradition must have some general bearing on modern attempts to reconstruct the economic régime of the Greek Dark Age. For the end of the Mycenaean world is a paradigm instance of civilisational collapse and, as in most such cases (Renfrew 1979:482-87), the ensuing age has been described and studied in a largely negative and contrastive sense, i.e. in terms that emphasize the *loss* of many of the general characteristics of the polities that preceded it – literacy, centralised administration, palaces, élite commodity production, etc. Thus, if the Mycenaean economy rested on the fruits of the labours of sedentary agriculturalists, it seems only 'natural' to suppose that the Dark Ages witnessed a 'reversion' to 'simpler' modes of subsistence, which (merely by structural opposition to settled farming) can only mean mobile, animal-based systems. The invasion of the Mycenaean polities by Dorians, if we allow them any historical reality at all, fit the ideological scheme nicely: the destruction of a sedentary, agricultural high civilisation in the south (= 'good') by nomadic, presumably pastoralist, culturally impoverished invaders from the north (= 'bad'). Aristotle (*Pol.* 1256a) would have applauded such value judgements. Surely modern archaeologists and historians of the Greek Dark Ages can do a little better by substituting model-building and empirical investigation for the ideology of the pastoral nomad?

Acknowledgements

Paul Halstead should have written this paper. Since he didn't, I hope that he will forgive the extent to which I have drawn on his work, and that I have not misrepresented it too badly; his criticisms of an earlier draft were invaluable. I gratefully acknowledge many helpful comments on the manuscript from Sue Alcock, Paul Cartledge, Clive Gamble, Steve Hodkinson, Bernie Knapp, Anthony Snodgrass, Peter Warren, and from my colleagues on the Nemea Valley Archaeological Project, Jack Davis, Jim Wright, and Jerry Rutter.

FITZWILLIAM COLLEGE, CAMBRIDGE JOHN F. CHERRY

REFERENCES

Angioni, G. 1975. 'I pastori dell'Ogliastra: Descrizione documentaria dell'ergologia tradizionale', *Studi Sardi* 23:355-442.

Aurenche, O. (ed.). 1984. *Nomades et sédentaires: perspectives ethnoarchéologiques.*

Barker, G.W.W. 1975. 'Prehistoric territories and economies in Central Italy', in E.S. Higgs (ed.), *Palaeoeconomy*, 111-75.

Barker, G.W.W. 1981. *Landscape and society: prehistoric central Italy.*

Barker, G.W.W. 1985. *Prehistoric farming in Europe.*

Baticle, Y. 1974. *L'élevage ovin dans les pays Européens de la Méditerranée occidentale.*

Barth, F. 1961. *Nomads of South Persia.*

Bennet, D.J.L. 1986. *Aspects of the administrative organisation of LMII-IIIB Crete: a study based on archaeological and textual data* (Ph.D. dissertation, University of Cambridge).

Bintliff, J.L. 1977. *Natural environment and human settlement in prehistoric Greece* (BAR Supplementary Series 28(i), 2 vols.).

Boesneck, J. and von den Driesch, A. 1981. 'Reste exotischer Tiere aus dem Heraion auf Samos', *AthMitt* 96: 245-48.

Bottema, S. 1974. *Late quaternary vegetation history of Northwestern Greece.*

Branigan, K. 1970. *The foundations of palatial Crete.*

Bray, W. (ed.). 1983. *Transhumance and pastoralism* (Special issue of *World Archaeology* 15(1)).

Braudel, F. 1972. *The Mediterranean and the Mediterranean world in the age of Philip II.*

Briant, P. 1982. *État et pasteur au Moyen-Orient ancien.*

Campbell, J.K. 1964. *Honour, family and patronage: a study of institutions and moral values in a Greek mountain community.*

Chang, C. 1981. *The archaeology of contemporary herding sites in Didyma, Greece* (Ph.D. dissertation, SUNY Binghampton).

Chang, C. 1984. 'The ethnoarchaeology of herding sites in Greece', *MASCA Journal* 3(2):44-48.

Chang, C. and Koster, H. A. 1986. 'Beyond bones: toward an archaeology of pastoralism', in M. B. Schiffer (ed.), *Advances in archaeological method and theory* 9: 97-148.

Cherry, J. F. 1981. 'Pattern and process in the earliest colonisation of the Mediterranean islands', *PPS* 47:41-68.

Cherry, J. F. 1983. 'Frogs round the pond: perspectives on current archaeological survey projects in the Mediterranean region', in D.R. Keller and D.W. Rupp (eds.), *Archaeological survey in the Mediterranean region* (BAR International Series 155) 375-416.

Cherry, J. F. 1984. 'The emergence of the state in the prehistoric Aegean', *PCPhS* 30:18-48.

Cherry, J. F. 1986. 'Polities and palaces: some problems in Minoan state formation', in C. Renfrew and J. F. Cherry (eds.), *Peer polity interaction and socio-political change* 19-45.

Cribb, R. L. 1982. *The archaeological dimensions of Near Eastern pastoralism* (Ph.D. dissertation, University of Southampton).

Cribb, R. L. 1983. 'On-site ethnoarchaeology', *Archaeological Review from Cambridge* 2:4-16.

Cribb, R. L. 1985. 'The analysis of ancient herding systems: an application of computer simulation in faunal studies', in G. Barker and C. Gamble (eds.), *Beyond domestication in prehistoric Europe* 75-103.

Cristofini, B. *et al.* 1979. 'Utilisation des parcours par les éleveurs de la Castagniccia', in G. Molénat and R. Jarrige (ed.), *Utilisation par les ruminants des pâturages d'altitude et parcours méditerranéens* 409-25.

Cullen, T. 1985. 'Social implications of ceramic style in the Neolithic Peloponnese', in W.D. Kingery (ed.), *Ancient technology to modern science* 77-98.

Dahl, G. and Hjort, A. 1976. *Having herds: pastoral herd growth and household economy.*

Davidson, I. 1980. 'Transhumance, Spain and ethnoarchaeology', *Antiquity* 54:144-47.

Davis, J. 1977. *People of the Mediterranean: an essay in comparative social anthropology.*

Dyson-Hudson, R. and Dyson-Hudson, N. 1980. 'Nomadic pastoralism', *Annual Review of Anthropology* 9:15-61.

Flannery, K.V. 1969. 'Origins and ecological effects of early Near Eastern domestication', in P.J. Ucko and G.W. Dimbleby (eds.), *The domestication of plants and animals* 73-100.

Fotiades, M. 1980. 'Transhumance: was it indeed practiced in the prehistoric Mediterranean?' [Abstract of a paper delivered at the 81st meeting of the Archaeological Institute of America] *AJA* 84:207.

Galvin, K.F. 1987. 'Forms of finance and forms of production: the evolution of specialised livestock production in the ancient Near East', in E.M. Brumfiel and T.K. Earle (eds.), *Specialisation, exchange and complex societies* 119-29.

Gamble, C.S. 1982. 'Animal husbandry, population and urbanisation, in C. Renfrew and M. Wagstaff (eds.), *An island polity: the archaeology of exploitation in Melos* 161-71.

Garnsey, P. 1988. 'Mountain economies in Southern Europe. Thoughts on the early history, continuity and individuality of Mediterranean upland pastoralism', originally in M. Mattmüller (ed.), *Itinera* 5/6 (1986), 7-29. Reproduced in this volume, pp. 196-209.

Geddes, D.S. 1983. 'Neolithic transhumance in the Mediterranean Pyrenees', *World Archaeology* 15:51-66.

Georgoudi, S. 1974. 'Quelques problèmes de la transhumance dans la Grèce ancienne', *REG* 87:155-85.

Gifford, D.P. 1978. 'Ethnoarchaeological observations of natural processes affecting cultural materials', in R.A. Gould (ed.), *Explorations in ethnoarchaeology* 77-102.

Gilman, A. 1981. 'The development of social stratification in bronze age Europe', *Current Anthropology* 22:1-8.

Halstead, P. L. J. 1977. 'Prehistoric Thessaly: the submergence of civilisation', in J.L. Bintliff (ed.), *Mycenaean geography* 23-29.

Halstead, P. L. J. 1981. 'Counting sheep in neolithic and bronze age Greece', in I. Hodder, G. Isaac, and N. Hammond (eds.), *Pattern of the past: studies in honour of David Clarke* 307-39.

Halstead, P. L. J. 1984. *Strategies for survival: an ecological approach to social and economic change in the early farming communities of Thessaly, N. Greece.* (Ph.D. dissertation, University of Cambridge).

Halstead, P. L. J. 1985. 'Man and other animals in later Greek prehistory'. Paper delivered to the 87th General Meeting of the Archaeological Institute of America, Washington, D.C. (*BSA* in press).

Halstead, P. L. J. 1987. 'Traditional and ancient rural economy in Mediterranean Europe: Plus ça change?', *JHS* 107:77-87.

Halstead, P. L. J. and O'Shea, J. 1982. 'A friend in need is a friend indeed', in C. Renfrew and S. Shennan (eds.), *Ranking, resource and exchange* 92-99.

Hammond, N. G. L. 1976. *Migrations and invasions in Greece and adjacent areas.*

Higgs, E. S. 1976. 'The history of European agriculture: the uplands', *Philosophical Transactions of the Royal Society of London* B 275:159-73.

Higgs, E. S. and Jarman, M. R. 1972. 'The origins of animal and plant husbandry', in E.S. Higgs (ed.), *Papers in economic prehistory* 3-13.

Higgs, E. S., Vita-Finzi, C., Harris, D. R. and Fagg, A. E. 1967. 'The climate, environment and industries of stone age Greece: Part III', *PPS* 33:1-29.

Higham, C. F. W. 1967. 'The economy of iron age Vejleby (Denmark)', *Acta Archaeologica* 38:222-41.

Höeg, C. 1925. *Les Saracatsans* (3 vols.).

Hole, F. D. 1974. 'Tepe Tula'i: an early campsite in Khuzistan, Iran', *Palaeorient* 2:219-42.

Hole, F. D. 1978. 'Pastoral nomadism in western Iran', in R. A. Gould (ed.), *Explorations in ethnoarchaeology* 127-79.

Hole, F. D. 1979. 'Rediscovering the past in the present: ethnoarchaeology in Luristan, Iran', in C. Kramer (ed.), *Ethnoarchaeology: implications of ethnography for archaeology* 192-218.

Irons, W. and Dyson-Hudson, N. (eds.) 1972. *Perspectives on nomadism.*

Jacobsen, T. W. 1978, 'Transhumance as a mechanism of exchange in Neolithic Greece', *Archaeological Institute of America, Abstracts* 3:47.

Jacobsen, T. W. 1984. 'Seasonal pastoralism in the Neolithic of Southern Greece: a consideration of the ecology of Neolithic Urfirnis pottery', in P. M. Rice (ed.), *Pots and potters: current approaches in ceramic archaeology* 27-43.

Jarman, M. R. 1972. 'Appendix VI: The fauna', in P. M. Warren, *Myrtos* 318-20.

Jarman, M. R. and Jarman, H.N. 1968. 'The fauna and economy of early Neolithic Knossos', *BSA* 63:241-64.

Jost, M. 1975. 'Statuettes de bronze provenant de Lykosoura', *BCH* 99:339-64.

Kavadias, G. B. 1965. *Pasteurs-nomades méditerranéens: les Saracatsans de Grèce.*

Khazanov, A.M. 1984. *Nomads and the outside world.*

Killen, J.T. 1964. 'The wool industry of Crete in the Late Bronze Age', *BSA* 59:1-15.

Killen, J.T. 1985. 'The Linear B tablets and the Mycenaean economy', in A. Morpurgo-Davies and Y. Duhoux (eds.), *Linear B: A 1984 Survey* 241-305.

Kirsten, E. 1983. 'Gebirgshirtentum und Sesshaftigkeit. Die Bedeutung der Dark Ages fur die griechische Staatenwelt: Doris und Sparta', in S. Deger-Jalkotzy (ed.). *Öst. Akad. der Wiss. Phil-Hist. Kl. Sitzungsberichte 418 Band.*

Klein, J. 1920. *The Mesta: a study of Spanish economic history.*

Koster, H.A. 1976. 'The thousand year road', *Expedition* 19(1):19-28.

Koster, H.A. 1977. *The ecology of pastoralism in relation to changing patterns of land use in the Northeast Peloponnese* (Ph.D. dissertation, University of Pennsylvania).

Koster, H.A. and Koster, J.B. 1976. 'Competition or symbiosis? Pastoral adaptive strategies in the Southern Argolid, Greece', in M. Dimen and E. Friedel (eds.), *Regional variation in modern Greece and Cyprus: toward a perspective on the ethnography of Greece* (Annals of the New York Academy of Sciences 268) 275-85.
Lees, S.H. and Bates, D.G. 1974. 'The origins of specialised nomadic pastoralism: a systematic model', *American Antiquity* 39:187-93.
Lewthwaite, J. 1981. 'Plain tails from the hills: transhumance in Mediterranean archaeology', in A. Sheridan and G. Bailey (eds.), *Economic Archaeology* (BAR International Series 96) 57-66.
Lewthwaite, J. 1982. 'Acorns for the ancestors: the prehistoric exploitation of woodland in the West Mediterranean', in M. Bell and S. Limbrey (eds.), *Archaeological aspects of woodland ecology* (BAR International Series 146) 217-30.
Lewthwaite, J. 1984a. Pastore, Padrone: the social dimensions of pastoralism in prenuragic Sardinia', in W. H. Waldren, R. Chapman, J. Lewthwaite and R.-C. Kennard (eds.), *The Deya conference of prehistory: early settlement in the Western Mediterranean islands and their peripheral areas* (BAR International Series 229(i)), 251-68.
Lewthwaite, J. 1984b. 'The Art of Corse Herding: archaeological insights from recent pastoral practices on West Mediterranean islands', in J. Clutton-Brock and C. Grigson (eds.), *Animals and archaeology 3: early herders and their flocks* (BAR International Series 202) 25-37.
Lilliu, G. 1975. *La Civiltà dei Sardi.*
Müller, E. 1938. 'Die Herdenwanderungen im Mittelmeergebiet', *Petermanns Geographische Mitteilungen* 84:384ff.
Murray, P. and Chang, C. 1981. 'An ethnoarchaeological study of a contemporary herder's site', *JFA* 8:372-81.
Pasquinucci, M. 1979. 'La transumanza nell'Italia Romana', in E. Gabba and M. Pasquinucci (eds.), *Strutture agrarie e allevamento transumante nell'Italia Romana* 79-182.
Payne, S. 1973. 'Kill-off patterns of sheep and goats: the mandibles from Aşvan Kale', *Anatolian Studies* 23:281-303.
Payne, S. 1975. 'Faunal change at Franchthi Cave from 20,000 BC to 3,000 BC', in A. T. Clason (ed.), *Archaeozoological Studies* 120-31.
Payne, S. 1985. 'Zoo-archaeology in Greece: a reader's guide', in N.C. Wilkie and W.D.E. Coulson (eds.), *Contribution to Aegean archaeology: studies in honor of William A. McDonald* 211-44.
Prag, K. 1985. 'Ancient and modern pastoral migration in the Levant', *Levant* 17:81-88.
Quezel, P. 1967. 'La vegetation des hauts sommets du Pinde et de l'Olympe de Thessalie', *Vegetatio* 14:127-228.
Redding, R. W. 1981. *Decision making in subsistence herding of sheep and goats in the Middle East.* (Ph.D. thesis, University of Michigan).
Renfrew, C. 1972. *The emergence of civilisation.*
Renfrew, C. 1979. 'Systems collapse as social transformation: catastrophe and anastrophe in early state societies', in C. Renfrew and K.·L. Cooke (eds.), *Transformations: mathematical approaches to culture change* 481-506.
Robbins, L. H. 1973. 'Turkana material culture viewed from an archaeological perspective', *World Archaeology* 5:209-14.
Rutkowski, B. 1972. *Cult places in the Aegean world.*
Ryder, M.L. 1983. *Sheep and man.*
Semple, E.C. 1922. 'The influence of geographic conditions upon ancient Mediterranean stock-raising', *Annals of the Association of American Geographers* 12:3-38.
Shaw, B.D. 1982-3. '"Eaters of flesh, drinkers of milk": the ancient Mediterranean ideology of the pastoral nomad', *Ancient Society* 13-14:5-31.
Sherratt, A.G. 1981. 'Plough and pastoralism: aspects of the secondary products revolution', in I. Hodder, G. Isaac, and N. Hammond (eds.), *Pattern of the past: studies in honour of David Clarke* 261-305.
Sherratt, A.G. 1983. 'The secondary exploitation of animals in the Old World', *World Archaeology* 15:90-104.
Skydsgaard, J.E. 1974. *Transhumance in ancient Italy* (Analecta Romana Instituti Danici VII).
Sloan, R.E. and Duncan, M.A. 1978. 'Zooarchaeology of Nichoria', in G. Rapp and S.E. Aschenbrenner (eds.), *Excavations at Nichoria in Southwest Greece* 60-77.

Snodgrass, A.M. 1971. *The dark age of Greece.*
Snodgrass, A.M. 1980. *Archaic Greece: the age of experiment.*
Snodgrass, A.M. 1987. 'An experimental synthesis: the Early Iron Age of Greece' (Sather Lecture, University of California at Berkeley) in A.M. Snodgrass, *An Archaeology of Greece: the present state and future scope of a discipline* Ch. 6.
Solecki, R. 1979. 'Contemporary Kurdish winter-time inhabitants of Shanidar Cave, Iraq', *World Archaeology* 10:318-30.
Spooner, B. 1973. *The cultural anthropology of pastoral nomads* (*Addison-Wesley Modules in Anthropology* 5).
Sprengel, U. 1971. *Die Wanderherdenwirtschaft im mittel- und sudöstitalienischen Raum.*
Sterud, E.L. 1978. 'Prehistoric populations of the Dinaric Alps: an investigation of interregional interaction', in C.L. Redman *et al.* (eds.), *Social archaeology: beyond subsistence and dating* 381-408.
Turner, J. 1978. 'The vegetation of Greece during prehistoric times: the palynological evidence', in C. Doumas (ed.), *Thera and the Aegean world* 765-73.
Turner, J. and Greig, J. 1975. 'Some holocene pollen diagrams from Greece', *Review of Palaeobotany and Palynology* 20:171-204.
Turrill, W.B. 1929. *The plant-life of the Balkan peninsula.*
Ventris, M. and Chadwick, J. 1973. *Documents in Mycenaean Greek* (2nd ed.).
Vickery, K.V. 1936. *Food in early Greece* (Illinois Studies in the Social Sciences XX.3).
Wace, A.J.B. and Thompson, M.S. 1914. *The nomads of the Balkans.*
Wagstaff, M. and Gamble, C. 1982. 'Island resources and their limitations', in C. Renfrew and M. Wagstaff (eds.), *An island polity: the archaeology of exploitation in Melos* 95-105.
Walker, M.J. 1983. 'Laying a mega-myth: dolmens and drovers in prehistoric Spain', *World Archaeology* 15:37-50.
Warren, P.M. 1968. 'A textile town – 4500 years ago?', *Illustrated London News* February 7, 1968:25-30.
Warren, P.M. 1972. *Myrtos: an Early Bronze Age settlement in Crete.*
Warren, P.M. and Tzedhakis, J. 1974 'Debla, an Early Minoan settlement in Western Crete', *BSA* 69:299-342.
Warren, P.M. 1986 Review of C. Renfrew (ed.), *The archaeology of cult, Antiquity* 60:155-56.
Watrous, L.V. 1977. 'Aegean settlements and transhumance', *Temple University Aegean Symposium* 2:2-6.
Wertime, T.A. 1983. 'The furnace vs. the goat: the pyrotechnological industries and Mediterranean deforestation in antiquity', *JFA* 10:445-52.
Whitelaw, T.M. 1983. 'The settlement at Fournou Korifi, Myrtos, and aspects of Early Minoan social organisation', in O. Krzyszkowska and L. Nixon (eds.), *Minoan society* 323-45.
Wiegand, G. 1894. *Die Aromunen.*
Winnifrith, T. 1987. *The Vlachs: a history of a Balkan people.*

ANIMAL HUSBANDRY IN THE GREEK POLIS

Although the role of animal husbandry within the political economy of the Greek *poleis* has long been acknowledged, it has rarely been the subject of systematic historical study. The most extended treatments this century have been written not by historians but by the Giessen school of agronomists (Hörnschemeyer 1929; Winkelstern 1933; Brendel 1934; Zeissig 1934). Now over 50 years old, these works, which focussed heavily upon the identification of ancient breeds and relied upon a limited range of literary sources (mainly Homer, Aristotle and the Italian agronomists), are largely outdated in their methods and approach. On the other hand, historical discussions, most frequently found either in general works on ancient Greek economics (e.g. Michell 1957, 59ff.; Bolkestein 1958, 3ff.; cf. Rostovtzeff 1941, II 1190f.), have generally been comparatively brief. Apart from cataloguing which animals were kept and in which regions, most of these studies, sometimes under the influence of geographical approaches (e.g. Semple 1922; 1932, 297ff.), have confined their attention to the environmental constraints upon animal husbandry, without serious consideration of its economic and social context. Similarly, discussions of the agricultural economies of individual Greek *poleis* (e.g. Roebuck 1945; Legon 1981, 88ff.) have generally limited their perspective to the evidence for livestock in the state in question, with little analysis of the wider issues involved.

Because of the lack of recent systematic study I have chosen to adopt a broad perspective on the subject of animal husbandry within the Greek *poleis* in the Archaic, Classical and Hellenistic periods. My aim is to explore some of the fundamental questions surrounding the role of livestock, in order to construct a provisional general framework within which a more detailed examination might subsequently be attempted. (This essay excludes, however, both the subject of animal husbandry in Homeric society and that of livestock and religious sacrifice; and I shall concentrate mainly upon mainland and Aegean Greece, with only passing attention to the world of the Greek colonies.) In constructing this framework it is also my aim to react against a dominant stereotype of ancient Greek animal husbandry which assumes a near-universal pattern of seasonal transhumance and to substitute a more nuanced and complex interpretation of the subject.

The questions to be addressed in this essay can be posed quite simply. Historians are generally agreed that Greek *polis* society was founded upon the practice of agriculture, which was the essential means of subsistence for the mass of ordinary citizens and an important source of wealth for the majority of the rich. What role did animal husbandry play within Greek agriculture vis-a-vis arable farming and arboriculture? And how important was it in the procurement of subsistence and surplus? To answer these questions is less easy, though preliminary orientation may be gained from a variety of sources. Such evidence as the number of *poleis* which put livestock motifs on their coinage (Plant 1979), the frequent attestations of the pillage of livestock in warfare (e.g. Ducrey 1968, 87ff.) or of inter-state conflicts concerning the possession of areas of pasturage (Sartre 1979, 214ff.; Jameson, forthcoming) and the casual listing of animals and pastoral items among the products of Attica and other regions (e.g. Athenaios, *Deipnosophistai* 27d-f, 28a, 43c, 131d, 282b, 402e, 618d-e, 640b-c, 658a-e, 691a) all suggests the prominence of animal husbandry throughout the Greek rural landscape.

This abundant evidence can be set against the absence for most of our period of any contemporary practical discussion of animal husbandry comparable with the treatments of arable culture in Hesiod's *Works and days* and Xenophon's *Oikonomikos* (Hopper 1979, 157; Rostovtzeff 1941, II 1183; III 1619). This phenomenon is not difficult to comprehend. First, most owners of livestock were primarily arable farmers. The lack of treatises on animal husbandry reflects the unimportance of specialized pastoralism, at least among those within the mainstream of *polis* life, in comparison with mixed arable and pastoral units. This is not to say that specialized pastoralism was non-existent (cf. Arist. *Pol.* 1319a19-23). Possible examples include the *synnoma* recorded from Rhodes, a group of men, corresponding to a phratry, whose name may suggest common possession of pasturage (Ehrenberg 1969, 11) and the village community (*chiliastys*) of the Peproioi, S.E. of Erythrai, whose eponymous magistrates in the fourth century were guardians of marshes used for pasturage (Buckler 1934). It is also conceivable that some of the dedicators of bronze statuettes of herdsmen and Hermes Kriophoros in S.W. Arkadia in the late archaic period (Lamb 1925/6; Jost 1975) may have been specialized pastoralists, owners in their own right rather than hired men serving arable farmers.

Most of this evidence concerns persons or places on the margins of *polis* society, contexts in which specialized pastoralism may have had a place. The bulk of the ancient evidence for the ownership of animals by citizens, however, relates to agro-pastoral units.[1] The vast majority of citizens were landowners who derived their primary subsistence and wealth from arable farming, cultivating grain as their staple crop. Consequently, the role and significance of animal husbandry within the Greek *poleis* must have been first and foremost the product of its interrelationship with arable culture. Within this context there was an ideological tendency to link arable farming with the cause of civilization and culture (e.g. Moschion, fr.6 Nauck ed.2; Xanthakis-Karamanos 1981) and, by contrast, to denigrate animal husbandry (Sallares 1986, ch.3. iv). A powerful image characterizing the state of affairs after the destruction of a *polis* was of its territory being abandoned to become a sheepwalk (Isokrates 14.31; Lykourgos, *Against Leokrates* 145). Herding was a much lower-status activity than arable farming and one for which larger landowners typically employed slaves or hired professional shepherds (see II below). Accordingly, it is probable that the detailed techniques of animal management were of less direct concern to, and considered a less suitable topic of conversation among, the social elite from whom almost all our literary evidence comes.

Despite the lack of extended commentary upon animal husbandry, the evidence suggests that livestock were an unexceptional part of many agricultural estates. For example, although there is little explicit discussion of domesticated livestock in Attic literature, the correct conclusion is not necessarily that animal husbandry was little developed (Hopper 1979, 157ff.). There were no doubt many estates without animals, as is suggested by the admittedly fragmentary evidence of the stelai recording the sale of confiscated possessions belonging to those accused of profaning the Mysteries and mutilating the Herms in 415/4. These make mention of animals, or of facilities for animals, on only two properties (Pritchett 1953, 265: V 39-40; 272, VI 68-73). Nevertheless, examples of combined agro-pastoral estates, such as those of Euktemon, Theophon and Stratokles (Isaios 6.33; 11.41, 43; cf. also [Demosthenes] 47.52), appear in the Attic orators; and other evidence suggests that they were far from uncommon. In Xenophon's *Memorabilia* Sokrates clearly assumes that Kriton, a wealthy Athenian, will own sheep; and the products from his estate which he put into store included wool, as well as grain, oil and wine (2.9.2, 4; cf. also the reference to men rich in the same items in the *Poroi* 5.3). The assumption that Athenian landowners were also likely to be owners of livestock runs throughout not only the *Memorabilia* (2.3.9, 7.13-14; 3.11.4; 4.3.10) but also, significantly, the *Oikonomikos* (1.9; 5.3, 6, 18, 20; 7.20; 20.23), despite its restriction of systematic discussion to arable farming and arboriculture. It is the *Oikonomikos* itself (5.3) which provides the revealing statement that the *technē* of animal husbandry (*probateutikē*) is linked with that of tillage (*geōrgia*). That the agro-pastoral farm was not merely an upper-class phenomenon in Attica is confirmed by Thucydides' remarks (2.14; 7.25.5) concerning the *probata* of the mass of Athenian farmers at the time of the Peloponnesian invasions of 431 and 413. Note in addition the fictional example of the ordinary farmer Strepsiades in Aristophanes, *Clouds* 45, 50, 71-2 (cf. also Eupolis fr.13, Edmonds 1957-61, I 318). In other places and at other times we find a similar pattern. Though Hesiod, like Xenophon, focusses upon arable culture, he also assumes the practice of animal husbandry (*Works and days* 230-7, 308, 536-46, 588-92, 775, 786-7, 795-7; cf. *Theogony* 22-3). The *Homeric hymn to Earth* (30), the estate which Xenophon dedicated to the goddess Artemis at Skillous in Elis (*Anabasis* 5.3.9-11), the *Greek anthology* (6.31; 7.740), the *tabula Heracleensis* from southern Italy (*IG* XI 645, i 138ff.) and the inventories of agricultural structures and trees available for the use of private lessees of the temple estates of Delian Apollo (Kent 1948, 292ff.) all provide examples of the combination of arable and/or arboriculture with animal husbandry.

I *Animal husbandry on the agro-pastoral farm*

a) *Models of symbiosis and divorce*

Is it possible to be more specific about the role of animal husbandry on agro-pastoral estates? A crucial question here is to what extent the relationship between arable farming or arboriculture and animal husbandry was one of true symbiosis or simply of coexistence without direct interaction. As a starting-point for discussion it can be noted that recent work (e.g. Halstead 1987; Cherry, in this volume) has highlighted a dichotomy between two broadly conflicting models employed explicitly or implicitly in studies of Mediterranean agriculture. The first, and older, model, which draws its inspiration from observations of traditional patterns of agriculture in recent centuries, assumes a large measure of divorce between arable and pastoral elements. On this view the standard practice in arable culture was the alternation of cereal crops with fallow ploughed bare in the spring. This practice reduced the availability of lowland summer grazing, thereby enforcing the seasonal transhumance of sheep and goats which was in any case a basic feature of animal husbandry owing to the conditioning factors of climate and relief which created an arid lowland summer environment alongside a temperate highland one. From spring to autumn stock were maintained upon mountain pastures, the cultivated fields thus being deprived of their manure. For a large part of the present century this model became virtually a stereotype of ancient Greek agriculture through its repetition in several influential scholarly works (e.g. Semple 1922; 1932, 297ff.; Jardé 1925, 86ff.; Michell 1957, 38ff.; Robert 1949a; Georgoudi 1974). Recently, however, some studies have proposed an alternative model which questions the applicability of recent traditional agricultural practices to the ancient past and argues for a fair degree of symbiosis between arable culture and animal husbandry. It emphasizes the rotation of cereals and legumes in arable cultivation, with a consequent increased availability of summer fodder in the lowlands, thus evening out seasonal imbalances and diminishing the need for transhumance. In return livestock provided manure and consumed weeds, thereby maintaining soil fertility and crop productivity. This model has gained increasing ground in the field of Greek prehistory and elements of it can also be found in recent studies of the historical period (Halstead 1981, 328; 1984, 315ff.; cf. White 1970, 119ff.; Koster 1977, 247ff.; Garnsey 1988).

The operation of these alternative systems of cultivation and herding has been linked to wider social and economic factors. The employment of the first system in recent traditional farming has been associated with the practice of plough agriculture conducted from nucleated settlements at some distance from a scatter of dispersed and fragmented landholdings. In these conditions the extensive cultivation of distant fields through the biennial rotation of cereal crops and bare fallow was adopted as a labour-saving tactic, as part of which labour-intensive legumes were excluded from most of the fields and confined to garden plots nearest the home base. Furthermore, the impracticability of grazing animals on their own arable lands often led different households to pool their livestock into larger communal herds which itself necessitated the exploitation of more distant seasonal grazing. In contrast, the second system has been associated with the practice of labour-intensive hand cultivation of small, closely associated plots accompanied by a dispersed pattern of settlement close to the fields, conditions conducive to more complex rotation schemes and the maintenance of small herds on or near the arable lands.

One must obviously exercise cautious judgement concerning the extent to which either of these models in its simplest form can characterize the varying realities of ancient Greek agriculture throughout a range of different environments over a number of centuries. Nevertheless, the dichotomy between these models may provide a suitable framework within which we can begin to analyse the role of animal husbandry.[2] The first variable to consider is the method of cultivation. The literary sources from Hesiod (*W.D.* 405) onwards assume ploughing, normally with oxen, as the standard method. Aristotle (*Pol.* 1252b12), commenting on Hesiod's evidence, states that 'the ox is the poor man's slave'. Yet was plough agriculture quite so universal? Given the diversity of physical environments and social structures among the Greek *poleis*, generalization is obviously hazardous. Calculations have, however, suggested that an average household needed to possess approximately 5 hectares (ha.) for the use of oxen to be feasible, partly because of the lower yields of plough agriculture compared with hand cultivation, partly because of the heavy maintenance costs of working oxen (White 1970, 336). There is some evidence to suggest that land allotments in the region of 5 ha. were regarded as the rule-of-thumb equivalent of a basic hoplite lot (Burford Cooper 1977/78, 168ff.) and no doubt some hoplites owned more than that. But in the conditions of high population relative to landed resources which existed during much of our period most *poleis* must also have contained many poorer households of sub-hoplite status who depended primarily upon farming (*ibid.* 170). Such households might have sustained themselves on plots perhaps as small as 2 ha. from the much higher yields of continuous (without biennial fallow) spade or hoe cultivation (White 1970, 345ff.; Halstead 1984, 315f.). Hesiod himself refers to the existence of the farmer with no oxen (*W.D.* 451), depicting him as someone who would ask to share another man's, but who might well be refused and have to do without. It may be that in later periods the existence of hand cultivation is generally obscured by the interest of the literary sources in citizens of hoplite status and above. Our period has no equivalent of the late first-century A.D. philosopher Dio Chrysostom, who, disdaining established cultural values, took pleasure in giving a detailed description of a small two-family farm, established by two ex-herdsmen out of marginal land near Cape Kaphereus in S.E. Euboia, which grew a mixture of wheat, barley, millet, beans and chickpeas, besides vines, fruit trees and vegetables, all cultivated entirely by hand (*Orations* 7.10-80, esp. 47). Hand cultivation is sometimes dictated additionally by the nature of the available arable land in areas of terrace farming (Forbes 1982; cf. Pliny, *H.N.* 18.178). Quite apart from its implications for the husbandry of other animals, the extent of the ownership of draught livestock (and of larger herds of breeding animals upon which smaller farmers in need of replacements may have been dependent) is an important issue in its own right as a determinant of differentiation among the citizen body.

Secondly, there is the issue of the distance between farm base and fields, which embraces both the question of the fragmentation of holdings and the currently controversial debate concerning the existence of nucleated or dispersed settlement. Fragmentation of landholdings is certainly evident in Attica at least (Davies 1981, 52-4; Osborne 1985a, ch.3), although it may not necessarily have occurred to the extreme extent often found in modern Greece (Thompson 1963), owing to the permissibility of endogamy and brother-and-sister exchange marriages explicitly prohibited by the rules of the Orthodox Church and the civil code of the modern Greek state (Philo, *On special laws* 3.4.22; Osborne 1985a, 135ff.; Friedl 1963, 64f.). Fragmentation could also be minimized in other ways (Lane Fox 1985, 216ff.). The traditional interpretation of settlement patterns among citizens of the Greek *poleis* is that nucleated settlements in town or villages were the norm with a minimum of dispersed farmsteads (cf. Pečirka 1973, 115ff.), a conclusion strengthened by two recent studies by Osborne, one (1985a, ch.2) regarding the overall settlement pattern in classical Attica, the other concerning farm buildings, usually the property of wealthy men, attested in epigraphic evidence from various *poleis* (1985b). His studies do, however, highlight both the contrast between the Attic pattern and those prevailing in several *poleis* in the Black Sea region and in Magna Graecia (1985a, 37-8) and the frequency even in mainland and Aegean Greece of permanent individual rural structures, a phenomenon already highlighted by Pečirka and now being documented in detail by a number of intensive archaeological surveys (Keller & Rupp 1983; Bintliff & Snodgrass 1985, 139). Even if many of these structures are not inhabited farmsteads, they are likely to be seasonal agricultural bases, such as those mentioned by Herodotus (1.17) in the territory of late seventh-century Miletos. This would in itself suggest a labour-intensive pattern of land use. This hypothesis is reinforced by the background scatter of ancient pottery throughout the agricultural landscape frequently detected by archaeological survey which probably reflects the intensive spreading of manure (Gallant 1983; Bintliff & Snodgrass 1985; Halstead 1987).

A further variable relevant to the issues under consideration is the nature of the agricultural labour force, since as Pečirka (1973, 118f.) has reminded us, 'the lower the status of the workers, the more possible or likely their residence on the farm'. Here we encounter a factor not applicable to either the recent traditional societies or the small-scale prehistoric communities whose study has generated the models of agricultural exploitation considered above – namely, the penetration of chattel-slavery or other forms of servile labour throughout many levels of ancient Greek agriculture, especially from the late archaic period onwards (Ste. Croix 1957; 1981; Pippidi 1973; Jameson 1977/78; cf., however, the recent challenge to this view by Wood 1983). Jameson has suggested that slave labour was important in permitting forms of intensification on Athenian citizen estates which would have been impossible with family labour alone. If this is correct, the use of slaves may have counteracted the potential effects of cultivation by plough oxen, nucleated settlements and fragmented holdings which, along the lines of the first model above, would normally be associated with the practice of extensive cultivation. This may explain the indications of intensive farming revealed by archaeological survey. Note, for instance, the image of the farm slave in Menander's *Dyskolos* (414ff.), with hoe in hand and orders to dig on the farm. Although Osborne (1985a, 143) has suggested that the use of slaves did not extend to the poorest farming households, this would, as was seen above, have been the group most likely and able to practise intensive hand cultivation with family labour alone, owing to the small size of their holdings.

b) *Cropping practices and the provision of grazing and fodder*

This brief survey of some of the factors conditioning patterns of agricultural exploitation is sufficient to indicate a complex variety of circumstances, many of which do not coincide exactly with the broad patterns predicated in the two models above. This becomes still more clear from the variety of cropping practices in operation, each with a different potential for the extent of integration with animal husbandry. The alternation of the cereal crop with short fallow is referred to with approval by several agricultural writers, who recommend frequent tillage of the fallow, starting in the spring, if not before, and continuing into the summer (Hesiod, *W.D.* 461-4; Xenophon, *Oik.* 16.10-15; Theophrastos, *C.P.* 3.20.8; cf. 25, 26; Jardé 1925, 21ff.; Jameson 1977/78, 127). Here we see the use of more labour-intensive methods of cultivation to reduce rather than increase the potential for symbiosis with animal husbandry as was postulated in the second model cited above. This is especially obvious on the estate of Ischomachos in the *Oikonomikos* where the fallow is not merely ploughed but may also be dug by the slaves. Although, as we saw above, Xenophon assumes the ownership of livestock, there appears to be a large measure of divorce between animal husbandry and arable cultivation on this estate. This is evidenced by the stress upon the value of the herbage as manure (cf. also 20.11-12) which suggests that animals were not permitted to graze it during the winter.

Ischomachos' methods, however, stand at one end of the range of options and need not be taken as typical. Here prohibitions in land leases are instructive in indicating some of the varied practices which tenants, usually wealthy men, would otherwise be expected to employ. On the one hand, they indicate that fallowing was abandoned by some farmers in favour of annual cereal-cropping (Jameson 1977/78, 127), a practice which would have the ambiguous effect of reducing winter pasturage on arable land but doubling the availability of late-summer stubble-grazing. On the other hand, they suggest that pasturing livestock on green fallow, a procedure also mentioned by Theokritos (*Idylls* 25.99), was often practised (Michell 1957, 54). In this latter case, even with spring ploughing of the fallow, the combination of stubble and winter fallow could between them have provided adequate grazing for modest flocks for much of the year. Under this régime the difficult periods of the year would have been after the spring ploughing and after the exhaustion of the stubble in the autumn. Since, however, spring and autumn are the two seasons of natural plant growth in the Mediterranean, these temporary shortages could perhaps often have been remedied from local non-arable resources where there were sufficient areas of suitable wild vegetation nearby (Lewthwaite 1981, 63). This is a situation in which the alternative models of cultivation and herding specified above can in reality be seen to overlap, since even under a régime of biennial fallow ploughed bare in the spring there remained opportunities for the rearing of livestock in symbiosis with arable culture. Moreover, if one follows Jameson's suggestion (1977/78, 127) that the exhortations of agricultural writers 'can ... be taken to suggest that the benefits were not always apparent or even real to all farmers', one wonders how often the spring ploughing and repeated tilling of the fallow which they recommend was honoured in the breach rather than in the observance, a point which Ischomachos himself appears to concede (*Oik*. 20.3).

There is in addition plenty of evidence for other cropping practices besides cereal-fallow rotation which were conducive to the integration of arable culture with animal husbandry. Book 8 of Theophrastos' *Enquiry into Plants* (*H.P.*) displays a knowledge of a range of ten different annually-sown pulses (in addition to other non-cereal crops like sesame and *erysimon*), at least four of which were grown in more than one variety (8.5.1). The fact that he treats them in the same book as his discussion of cereals, and separate from his examination of garden vegetables and herbs in Book 7, suggests that they were viewed as a normal component of field agriculture used in rotation with cereal crops and not just confined to small-scale garden horticulture.[3] The same impression is given by his remark (8.2.8) that they were not abundant on Melos despite the productivity of the soil, as if normally the opposite was to be expected. Elsewhere (*C.P.* 3.20.7) he recommends the choice of early pulses in order to leave the land free in summer for preparatory work for the following year's crop. Theophrastos' evidence is corroborated by other sources. At least 9 of his 10 pulses are also mentioned by other writers from the fourth century or earlier.[4] As Jardé (1925, 84ff.) pointed out long ago, though he did not pursue its implications, the terms of certain Attic leases, which allow tenants to cultivate the land as they wish, except in the final year when half is to be handed over as fallow, imply a variety of practices other than cereal-fallow rotation. A cereal-pulse rotation, with a free choice of which pulse(s), is even specified in at least two leases (*IG* II 1241. 21-4; 2493. 8-10). This is not surprising, since the ancient Greeks shared the widespread modern notion that pulses reinvigorate the soil (n.7 below).

Although proper estimation of the extent of the cultivation of pulses as field crops would require detailed research beyond the scope of this paper, the above evidence is sufficient to suggest that it was an important agricultural practice. This impression is confirmed by such episodes as the famine at Ainos on the Thracian coast mentioned in the Hippocratic corpus (*Epidemics* 2.4.3; 6.4.11), which records the illnesses which befell those people compelled to rely upon repeated consumption of pulses – especially *Lathyrus sativus*, chickling or grass pea (Grmek 1983, 324-7) – and bitter vetch.[5] The very availability of such cereal alternatives indicates the significant scale of their cultivation. Indeed, the use of *lathyros* as a common foodstuff is assumed by the fourth-century comic poets Anaxandrides (fr. 41.43, Edmonds 1957-61, II 61) and Alexis (fr. 162.12, Edmonds *ibid.*, II 450). It was only its excessive or exclusive consumption at a time of nutritional deficiency which led to medical problems (Grmek 1983, 326-7; cf. Brouk 1975, 124-5). Similarly, bitter vetch was at Athens at different times both a valued component of the life of ease conjured up by the chorus of farmers in Aristophanes' *Peace* (1136) and an emergency staple, reliance upon which was for Demosthenes (22.15) the symbol of the *polis'* desperate condition during the wars of the late 370s.[6]

In addition to such fluctuations in the level of human consumption, there were doubtless also considerable temporal and geographical variations in the extent of cultivation of the various types of pulse. It has been suggested, for example, that dry climatic conditions may have restricted the extent of pulse cultivation in ancient Attica, as in modern times, compared with cooler and wetter areas (Sallares 1986, ch.3, ii). Certainly it is necessary to take account of such factors as the wide agroclimatological range of environments in which the Greek *poleis* were situated. In general, however, the evidence suggests that pulse cultivation was a significant part of the agriculture of the *polis*. This was true even in the case of the broad bean or horsebean (*Vicia faba*), despite the fact that not only is it the least drought-resistant pulse crop (Arnon 1972, II 235), but in addition its ingestion precipitates in certain individuals with a particular genetic defect the disease called favism, which seems to have been widespread in Greek antiquity (the following is heavily dependent upon Grmek 1983, 307-54). Despite the Pythagoreans' probable awareness of the link between broad beans and the disease, there is no evidence that their taboo against broad beans spread outside their sect or led any other communities to abandon or restrict their cultivation or consumption. On the contrary, the evidence of medical texts, the numerous references to fields of broad beans in stories concerning the Pythagoreans and, indeed, the continued existence of the disease itself all testify in the opposite direction. The rationalist mentality of Hippocratic and later medical writers prevented them from understanding or accepting the Pythagoreans' semi-magical insight. Moreover, the relative infrequency of the disease itself compared with the large proportion of the population bearing the underlying genetic defect meant that considerable numbers of people were able to eat broad beans without ill effect. Consequently, the true nature of favism and its causes went unrealized before the start of modern investigations beginning in the mid-nineteenth century.

The significance of the evidence for the use of pulse crops in arable cultivation is partly as an indicator of intensive farming methods congruent with the integration of animal husbandry, partly that the crops themselves could provide nutrition for livestock at times of the year when cereal stubble or fallow was either unavailable or in short supply. Theophrastos himself clearly viewed the pulses he mentioned as intended for animal as well as human consumption (*H.P.* 8.3.5, 9.3; cf. 7.3), as does Aristotle (*H.A.* 596a25), a point supported by the evidence concerning Ainos and by their use as forage crops in Roman farming (White 1970, 214ff.). When grown directly for animal consumption, they could be grazed *in situ* or harvested for use as fodder at lean times of the year. Pulses utilized as a fodder or forage crop before they mature seed have indeed been shown to improve soil fertility and enhance subsequent winter cereal crops (Arnon 1972, I 465). Clearly the extent to which ancient farmers devoted such crops to animal rather than to human consumption will have varied considerably within our period; but even when consumed by humans the residues of the pulse crop would have been available for livestock.[7] The variety of pulses, their different times of sowing and ripening (Theophrastos, *H.P.* 8.1.2ff.) and their diverse nutritional uses all extended the potential for maintaining livestock side by side with arable culture.

This evidence in itself would lead us to question Jardé's assertions (1925, 88ff.) concerning the absence of non-cereal rotation crops and of fodder-cropping in ancient Greek agriculture. Theophrastos (*H.P.* 8.7.5, 9.2; *Characters* 4) also mentions the use of wheat, barley and *zeia*, a type of emmer, as fodder crops, the last explicitly for all animals. A significant addition to this range was lucerne, or alfalfa (*Medicago sativa*: Aristophanes, *Knights* 606; Arist. *H.A.* 522b27; 595b29; Theophr. *H.P.* 8.7.7; *C.P.* 2.15.6), first introduced to Greece in the early fifth-century Persian invasions and the subject of special treatment by the Hellenistic writer, Amphilochos of Athens, because of its importance in animal nutrition (Pliny, *H.N.* 18.144; White 1970, 191, 203, 217). Although of particular value as feed for horses, it was used for all kinds of stock (cf. Diod. 2.43.6; Arist. *H.A.* 522b26; Theophr. *H.P.* 8.7.7; Columella, *De re rustica* 2.10.25). Amphilochos also devoted considerable attention to a shrub called *kytisos* (tree-medick = *Medicago arborea*) which he says was discovered on Kythnos and from there transplanted throughout the Cyclades and then to other parts of Greece. Its cultivation is indicated in the fifth century by Demokritos of Abdera (*apud* Pliny, *H.N.* 13.131) and its use as animal fodder is directly attested by Eupolis (fr. 14, Edmonds 1957-61, I 320), Aristotle (*H.A.* 522b27) and in Aischylides' account of sheep-rearing on Hellenistic Keos (see below). It too was recommended for all types of stock, especially for its milk-producing qualities (Pliny, *H.N.* 13.130-4; Columella, *De re rustica* 5.12; *De arboribus* 28). A number of writers from Theophrastos onwards also emphasize the importance of lupines grown as a green manure crop which could in addition provide winter animal fodder (Semple 1932, 417 with refs.; White 1970, 216f.).

One context in which the cultivation of fodder crops may have been especially significant was the olive grove (Foxhall, unpub. MS; Koster 1977, 72f., 315). Fodder crops tolerate better than do cereals the poorer soils and sloping ground on which olives were often grown and their sowing could be coordinated with the slack time in olive cultivation. The waste products of olive culture, prunings and presscake (which are also available from other forms of arboriculture and from viticulture) can also pay an important role in maintaining livestock (cf. Arist. *H.A.* 596b25-6), as they clearly did on the Elder Cato's olive plantation (Foxhall, unpub. MS). Theophrastos' discussion (*H.P.* 2.7.4) of various types of animal dung is conducted in the context of tree cultivation; and since the dung is distinguished from litter manure, the implication is that stock were often grazed in or near the orchards themselves. This kind of symbiosis may have underpinned the economies of states like second-century Pidasa in Caria, described by Rostovtzeff (1941, I 671) as being based on olives and livestock, and of several of the temple estates on Delos and Rheneia which show a marked concentration on the combination of livestock and vines (Kent 1948, 299ff.).

c) *Animals on or near the cultivated fields*

This evidence for the range of potential ways of integrating animal husbandry with arable cultivation can be corroborated by direct references to the pasturage of livestock on or near the cultivated fields at all times of the year. The pasturing of stock on green fallow has already been mentioned and the grazing of sprouting crops to prevent plants on good soils from running wildly to leaf is also attested (Theophrastos, *H.P.* 8.7.4; Michell 1957, 52). Particularly instructive is the account of sheep-rearing on his native Keos given by the Hellenistic writer Aischylides,[8] as reported by Aelian (*On the characteristics of animals* 16.32).

> Each of the farmers owns but few sheep, the reason being that the soil of Keos is exceedingly poor and has no pasture land. So they threw *kytisos* and fig-leaves and the fallen leaves of the olive to the flocks,[9] also husks of various kinds of pulse, and they even sow thistles among the crops, all of which afford excellent feeding for the sheep. And from them they obtain milk which when curdled provides the finest cheese; and the same writer says that it is called Kythnian and that it is sold at the rate of 90 drachmas a talent. And lambs are also produced which are of remarkable beauty and are sold not at the price of ordinary lambs but for a far more impressive figure.

This passage provides an excellent example of labour-intensive, integrated agro-pastoral landholdings on which fodder crops and the residues of pulse and tree cultivation were utilized for rearing animals. The passage, moreover, seems to indicate that such agro-pastoral units with land available for growing *kytisos*, figs and olives were not the preserve of wealthier estates but widespread throughout the population. The extent to which sheep-rearing was geared towards the commercially-orientated production of milk and suckling lamb, as in modern Greece, is worthy of note. It is also implied that, reliant upon fodder rather than natural grazing, flocks were kept close to the cultivated land and its adjacent settlements or rural establishments, as on modern Melos where most small livestock are kept within five minutes' walking-time of the farmer's house (Wagstaff & Augustson 1982, 122; Cherry, Davis & Mantzourani, forthcoming). The two-family arable farm described by Dio Chrysostom which operated closer to the margins of subsistence also kept a small on- or near-site complement of eight she-goats, a hornless cow with calf and pigs reared on acorns and barley.

Evidence regarding the upper end of the social scale comes from a speech in the Demosthenic corpus (47.52-3; cf. 76) and a passage in Polybius (9.17.6; cf. Walbank 1957-79, II 143), both of which mention examples of soft-fleeced sheep owned by leisured landowners – at fourth-century Athens and third-century Kynaitha in Arkadia respectively – being pastured close to their home base during the spring or summer months.[10] Polybius tells us that the owner of the flock at Kynaitha kept his animals near the town; and the flock belonging to the plaintiff in the speech of Demosthenes cannot have been far from Athens or Peiraieus, since he claims that the defendant Theophemos and his associates seized his animals and two slaves that were with them before proceeding to make further thefts at his farmhouse near the Hippodrome. Indeed, when the plaintiff states (76) that 'he went and seized my furniture, slaves and sheep instead of receiving payment; I have a farm by the Hippodrome, so that he did not have far to go', the clear implication is not only that the farm was near the urban centre,[11] but also that the flock was close by the farm – so close in fact that one of the plaintiff's neighbours knew of the flock's seizure and was able to inform the plaintiff's wife before the thefts from the farmhouse had been completed (57). The 'jacketed' sheep (that is, sheep given coverings, often of leather, to protect their fleeces) attested in Megara, Attica, Taras and Epirus (Diogenes Laertius 6.41; Varro, *De re rustica* 2.2.18-20) are another example of animals maintained near their farmsteads, since, probably like the sheep of Keos, they were kept in stalls and fed with lucerne, *kytisos* and controlled quantities of fig leaves, straw, grape skins and bran.

It was not only such special forms of stock which were to be found grazing in or near to the plains in spring and summer. Accounts of military expeditions, especially those in the Peloponnese and central Greece, commonly refer to either the capture of large numbers of livestock or the attempt of the defenders to prevent such an occurrence by sending their animals away to remote places (e.g. Thuc. 2.14; 7.25.5; Xen. *Hell.* 3.2.26; 4.5.5-8, 6.4-6; 7.5.14-15; Plb. 4.4.1, 29.6; Livy 27.32.7-9). Since the capacity of armies to conduct wide-ranging operations into remote areas was limited (cf. Hanson 1983), it is clear, even when the texts themselves do not state it explicitly, that many of these animals were normally pastured in the most frequented farming areas. It has been suggested that the protection of such livestock was one of the intended functions of the vast town circuit-walls constructed particularly in these regions from the fourth century onwards (Martin 1973, 110). Particularly instructive is some evidence in Polybius (4.3.8 4.1) concerning the location of livestock in the late 220s in one such *polis*, Messene, which had built a large enceinte in the fourth century. The bandits who were raiding the Messenians' livestock from their base in neighbouring Phigaleia first plundered flocks located on the border, then homesteads in the countryside and, finally, approached the town itself and with the aid of scaling-ladders stole the animals from a farmstead situated apparently within the town walls.

One factor which enabled livestock to be maintained in farming areas in summer was the availability in a number of locations of uncultivated pasturage near the cultivated fields. One such type was marsh and wet meadows by the edge of lakes and rivers, near perennial springs and in coastal plains. As Theophrastos states, 'all things that grow in such places may be eaten by oxen and *probata*' (*H.P.* 4.8.13; cf. Arist. *H.A.* 7.23.8). Summer, when the water table was at its lowest, was precisely the time when such areas were most accessible to livestock. In the midst of indisputable generalizations about the dryness of the Greek summer it is easy to forget the existence of environmental niches in which such wet summer pastures can survive even in S. Greece and it is certain that their number and extent were considerably greater in antiquity than today. This is most obvious in areas like Lake Kopais, now thoroughly drained, but once such an important, if variable, grazing resource (Theophrastos, *H.P.* 4.10-12) that in the Hellenistic period the *poleis* of Akraiphia, Kopai and Orchomenos were all able to grant the privilege of free pasturage as suitable recompense to creditors whose loans could not be repaid (*SEG* xxix 439; xxii 432; xxvii 63). Orchomenos had an official, the *nomōnas*, specially charged with the collection of the pasture tax. But it was not only exceptional locations like Kopais which could in antiquity provide summer grazing not available today. Evidence from the fifth to the third century indicates that the upland plain of Mantineia in E. Arkadia, for instance, possessed considerably more surface water in antiquity than at present (Hodkinson 1981, 267f.), a factor which may help explain the presence of all the Mantineians' livestock there during a Theban raid at harvest time in 362 (Xen. *Hell.* 7.5.14-15).[12] Similar locations to those just mentioned are attested in many parts of the ancient Greek world (cf. Semple 1922, 23ff.; Robert 1980, 12f.; Theokritos, *Idylls* 25.7-11; Paus. 1.32.5-7; 7.22.11, 23.1; *IG* II^2 2498.13), as well as in numerous travellers' and geographic accounts and on maps up until the nineteenth century. Only with the advent of deep wells, especially those with power-driven pumps, from the last century onwards have many Greek plains assumed their current predominantly dry aspect (T.H. van Andel, pers. comm.).

In addition one should not forget the availability of grazing on scrubland on the lower slopes of nearby hillsides and in less cultivable niches in the low plateau and gently-rolling hill country, characterized by Neogen or Flysch soils, which were favoured areas of farming in antiquity (Bintliff 1977, 87ff.; Van Wersch 1972). Recent research has affirmed the important and varied nutritive value for sheep, goats and even cattle of this 'complex, beautiful and resilient mosaic of vegetation' (Rackham 1983, esp. 347; cf. id. 1982; Koster 1977, 72f.; Lewthwaite 1982) so disdained by earlier generations of students. Similarly, scholars are also coming to appreciate the significant contribution of woodland browsing and other arboreal resources to the maintenance of ruminants, especially during the summer months (Lewthwaite 1982, esp. 219f.; Garnsey 1988, 23).

Although many of these non-arable areas may have been public land, some of them seem to have been in private hands. Tracts of scrubland appear in three of the nine cases cited by Davies (1981, 53-4) in which we have details of the private property-holdings of wealthy Athenians; in two of these three the scrubland is, significantly, within their own deme. The estate which Xenophon dedicated to Artemis at Skillous provides another example. Pigs, goats, cattle and horses were all reared on the estate itself, which was a mixture of cereal, vine and orchard land, in association with meadows and tree-covered hills. It retained sufficient grazing to cater in addition for the draught animals of those attending the annual festival there after the harvest (*Anab.* 5.3.7-13). Discussions of how to overcome the problems of waterlogged and salty soils (Xen. *Oik.* 20.12; Theophrastos *C.P.* 3.6.3) may be evidence for private ownership of coastal pasture areas which the diligent farmer now wanted to turn into arable. For wealthy men who owned insufficient non-arable grazing one possibility was to lease it from temples who owned tracts of land for their sacred herds, or from other public bodies. Several of the estates on Delos and Mykonos leased from the temple of Delian Apollo seem to have included hillside and other grazing land (Kent 1948, 299ff.) and coastal pasture is included in land leased in the 320s by the Attic demes of Aixone and Peiraieus (*IG* II[2] 1196. A7, B10-11; 2498.13). As with the location of privately-owned scrubland, it is significant that most lessees of land belonging to demes were members of the deme concerned (Whitehead 1986, 157-8).

The presence of animals on or near arable land is attested, finally, through references to the use of their dung. Greek literature is full of references to manure (*kopros*), to the existence of manure heaps (*kopriai*) and, through a variety of terms (*kophinos koprophoros, kopragōgeō, kopragōgos, kopridzō, koprisis, koprōsis* etc.; cf. the refs. in *LSJ* ed.9), to the processes of carrying manure and depositing it on the fields. The writings of Theophrastos, moreover, reveal a great deal of sophistication in the use of *kopros*. He remarked, for example, that different trees and crops required different kinds and strengths, and some none at all (*H.P.* 2.7.3-4; 8.7.7; *C.P.* 3.6.2). He advised differential applications according to the richness of the soil (*H.P.* 8.6.3; *C.P.* 3.9.2, 20.2), and noted varying practices concerning methods of manuring and the relative timing of manuring and sowing (*H.P.* 7.5.1-2; *C.P.* 3.6.1-2, 9.3). Theophrastos suggested that manured land got as much as twenty days' start on unmanured land (*H.P.* 8.7.7). Landowners clearly set great store by its use, as is indicated in a fourth-century lease of the estate of Zeus Temenites at Arkesine on Amorgos in which the tenant was obliged to deposit 200 *medimnoi* of manure on the land each year (*IG* XII 7, 62.40-4). Owens' recent study of the *koprologoi* (*kopros*-collectors) at Athens suggests that they were private entrepreneurs engaged in the profitable collection and storage of an item in demand from farmers (1983, 48); and arrangements are also made for the sale of *kopros* in an inscription from Tegea, dated c.400, setting out various regulations of the temple of Athena Alea (*IG* V 2, 3.28, with Buck 1955, 201). The sources of the manure were no doubt the flocks and herds grazing upon temple lands, the terms of which the inscription also specifies.

For our purposes the difficulties in interpreting much of the above evidence are twofold. First, the word *kopros* was variously used to refer not just to animal dung, but also to human excrement (e.g. Theophrastos, *H.P.* 2.7.4), green manure (e.g. Xen. *Oik.* 16.12) and vegetation rotted down into liquid manure (*ibid.* 20.11-12), as well as all kinds of waste material no doubt composted on manure heaps. This variable semantic usage was at one with the practical usage of farmers who utilized all types of organic fertilizers on their lands, with in some cases a preference for human excrement. Theophrastos (*H.P.* 2.7.4), for example, judged that the latter was the most pungent kind of dung and cited the work of the otherwise unknown Chartodras (possibly a conflation of Chaereas and Androtion: Einarson & Link 1976, xx), who had deemed it the best. In most brief references to the spreading of manure the proportion of animal dung is, consequently, unguessable. Secondly, as is indicated by the evidence for the *koprologoi* and the regulations at Tegea, manure used by farmers, even if largely or partly animal dung, need not always have come from their own animals. Despite these words of caution, dung from one's own animals was obviously important, as is shown by the remainder of Theophrastos' citation from 'Chartodras', who had ranked the dung of pigs, goats, sheep, oxen and beasts of burden (in that order) according to its quality. Theophrastos himself went on to distinguish different kinds of litter manure as well as various ways of applying it. Elsewhere (*H.P.* 7.5.1) he noted that the most commended form of manure was that mixed with litter. The distinctions drawn between different animal manures and litters clearly imply the use of dung from the farmer's own animals, not of a composite heap purchased from an outside source; and the careful attention to its varying quality suggests considerable experience of the use of a valuable asset.

* * * * *

So far I have concentrated largely upon the evidence for varying degrees of symbiosis of agriculture and stock-breeding and the means by which it could be achieved, to the exclusion of the question of the pasturage of livestock away from cultivated land and the seasonal transhumant movements to and from mountain pastures which have usually been seen as the hallmarks of ancient Greek animal husbandry. The reason is not necessarily the insignificance of these phenomena, but rather the need to view them in context. The foregoing discussion has argued that, contrary to the view sometimes implicitly adopted (Semple 1922; 1932, 297ff.; Michell 1957, 59), the factors of climate and relief did not of themselves exercise a determinant influence upon patterns of pasturage. Seasonal migration was only one of a variety of options open to owners of stock and in many socio-economic circumstances it made little sense. The labour-intensive husbandry of small numbers of sheep practised on Keos, which functioned independently of mountain pasture, was geared to the commercial production of superior quality cheese and lambs for sale at above average prices. The suburban-based, soft-fleeced sheep mentioned by Polybius and Demosthenes were probably also intensively reared (the latter flock was only 50 in number) for the production of high quality wool, as was certainly the case with the 'jacketed' sheep also mentioned above. Furthermore, it is improbable that the small numbers of animals of various sorts typically kept on hand-cultivated smallholdings necessitated seasonal migration, given the full usage of the range of agricultural waste products permitted by the complementary feeding habits of different livestock (Halstead 1984, 318). In all these circumstances, whether the animals were reared for commercial purposes or for subsistence, the common denominators of labour-intensity and smallness of numbers were the crucial factors.

It is of course correct to point out that the resulting degree of integration between agriculture and animal husbandry fell short of that first achieved during the Agricultural Revolution in Britain by the adoption of ley-arable rotation systems (Sallares, n.d.). These entailed the practice of arable-pasture rotation on a given set of fields in continuously alternating cycles each lasting a few years at a time, in contrast with ancient practices which invariably maintained a sharp distinction between cultivated lands and those designated as permanent pasture. Ancient Greek farming never attained these higher levels of integration. However, through the grazing of fallow and stubble, of pulse and other forage crops, through the growing of fodder crops and the use of the residues from pulse and tree cultivation, through the maintenance of animals near the farmstead and through the deposition of animal dung on the cultivated fields, lower levels of symbiosis were often achieved which had important implications for the nature of agriculture.

II *The character and limitations of transhumance*

It remains now to consider the evidence which might be adduced for the use of more distant pastures. The distinction between pastures on the mountain and in the plain appears in several sources (e.g. *Homeric hymn to Hermes* 491-2). But it is the famous passage of Sophokles (*Oedipus rex* 1132ff.) which provides the most vivid account of seasonal transhumance, with the Theban slave of King Laios and a Corinthian hired shepherd both pasturing their flocks on Mt. Kithairon from spring to autumn before each returning to his lowland pastoral base. Additional evidence for the use of mountain pastures appears in references to livestock pasturing on the slopes of particular mountains (e.g. Euripides, *Andromache* 1100-1: sheep on Mt. Parnassos). A number of inscriptions also record the presence of herdsmen, sometimes seemingly free men, sometimes slaves, in places characterized by Robert (1949a, 154) as 'les landes et les bois qui bornent ordinairement le territoire de la cité, isolés et silencieux'. These border areas were frequently the source of conflict between neighbouring *poleis* and sometimes it is clear that grazing rights were the point (or one of the points) at issue (Sartre 1979, 214f.; Jameson, forthcoming; cf. also Thuc. 5.42.1; Plato, *Republic* 373d-e). Sometimes conflicts took place over the possession of tiny islands with seasonal – usually winter – grazing potential (Sartre 1979, 215; Robert 1949b). Further evidence for the movement of animals into or through the territory of another *polis* comes in the form of the agreements between the Lokrian *poleis* Myania and Hypnia (*BCH* 89, 1965, 665ff.) and between Aigai and Olympos in Aeolis in N.W. Asia Minor (*Rivista di Filologia* 94, 1966, 290ff.). (*IG* V 2, 3, relating to the passage of *sacred* herds through the pasture grounds of the temple of Athena Alea at Tegea, is excluded from consideration here.) Another body of evidence which has sometimes been deemed relevant concerns the granting of the right of *epinomia* to citizens from other *poleis*, either through a mutual inter-state concession applying to all citizens of the other *polis*, as in the early third-century agreement between Hierapytna and Praisos on Crete (*I. Cret.* III 4.1; cf. also *I. Cret.* III 3.4, a second-century agreement between Hierapytna and Priansos), or through a unilateral grant to a specific individual, most frequently a *proxenos* (Marek 1984, 147ff.; *SEG* xxvii 63). Although the precise meaning of *epinomia* has long been a matter of dispute, the evidence that the privilege was sometimes also given to citizens of the granting *polis* (*SEG* xxix 439; xxii 432) indicates that it signifies the privilege of exemption from the tax which a citizen or non-citizen pasturing animals on public grazing land would normally have to pay (Hennig 1977, 130; Marek 1984, 148f.).

Much of the above evidence has been outlined at length by Georgoudi (1974), whose discussion focusses largely, however, upon the environmental and climatic influences upon transhumance rather than its social and economic context or relationship with arable culture. The initial points to clarify are the nature and scale of the animal movements. First, although some of the evidence clearly relates to seasonal movements to pastures at higher altitudes, this is not obvious in every case. The provisions of the Aigai-Olympos treaty, for example, do not specify the type of animal movements which they are concerned to regulate and might well have been designed to cope with a variety of circumstances, such as those necessitated by sale to another owner and, in particular, movements to market. In lines 7-9 of the inscription Olympos exempts certain types of animals from import tax and, as Georgoudi has pointed out (1974, 178), the common denominator is that these animals were not normally saleable. The apparent restriction of tax to saleable animals, along with the limitations placed (lines 1-4) upon the amount of clothing, cereals and wine which the shepherds could bring with them and the prohibition upon their engaging in any employment, may indicate that movements to market were the primary issue here.

Similarly, in the case of grants of *epinomia* to individual *proxenoi*, it is not necessarily clear what kinds of animal movements, if any, were taking place. *Proxenoi* were not necessarily resident in their own *poleis*. In the Hellenistic period, from which almost all the grants of *epinomia* come, it was quite normal for a *proxenos* to be resident in the granting *polis* or travelling around as, for example, an officer or artist (Marek 1984, Teil IV; also 387f. & pers. comm.). Only 13 of the 80 or so proxeny decrees which include the privilege of *epinomia* are informative about the recipient's residence. Two of these relate to doctors obviously living in the granting *polis* (*IG* IX 1^2, 104 from Elateia; *IG* IX 1^2, 3, 7501 from Amphissa, concerning a Macedonian from Hyrkania). Two others concern a travelling poet and poetess, from Hypata in Thessaly and Smyrna in Ionia respectively, both granted *epinomia* in Lamia (*IG* IX 2. 63; 62). Neither is likely to have used the privilege. Indeed, the wide range of privileges given to these two artists and the similar sequence in which they are listed suggest the use of some form of standard privilege-formula. In general, although the privilege of *epinomia* cannot be dismissed as purely honorific, equally it is rarely possible to assume in any particular case that the grant was directly related to the economic circumstances of the recipient, especially when it is just one among a long list of honours. In at least some cases it must be a question of a 'potential' privilege only whose future use was not impossible but not specifically envisaged (Marek, pers. comm.). In addition to these qualifications, several of the grants of *epinomia* are also dubious evidence for vertical transhumance in as much as they concern states in regions like Boiotia and Thessaly (e.g. Orchomenos, Akraiphia, Kopai, Pherai and Gonnoi) whose prime summer grazing areas were low-altitude marshy plains. Some of the border disputes over grazing territory were also concerned with this sort of area rather than mountain pasture (Sartre 1979, 214).

Even in cases where some form of vertical movement is likely or certain, it appears that the scale of movement was often relatively short-distance. Most of the seasonal movements attested by disputes over mountain borders necessarily fall into this category, since few *poleis* possessed very extensive territories. Sometimes one suspects that the distance between home agricultural base and mountain pasture may have been very small in certain topographic situations. One of Theokritos' *Idylls* (8.1-2, 15-16) has a certain Menalkas herding sheep on the mountains but returning to his parents at nightfall. Koster's study of the modern agro-pastoral Arvanites in the Didyma basin in the Argolid peninsular indicates the possibility in some areas of exploiting highland pasture while remaining within a few hours of the farm base, sufficiently close for frequent journeys to maintain arable culture (1977, 74, 414ff.). One wonders whether the find-spots of the herdsmen inscriptions recorded by Robert, although nowadays remote in relation to modern settlements, were in antiquity really as isolated as he suggests from areas of cultivation and human activity. For example, the N. slopes of Ag. Ilias on Aigina where the stone commemorating the herdsman Koxios was reputedly found (Robert 1949a, 154), although three hours distant from the town of Aigina, are even today less than an hour's walk from cultivated land in the region of Pakhia Rakhi (personal observation in 1982 and 1983). The flocks herded by Koxios could well have been associated with an arable estate in this area. (This example highlights the need for precise, preferably first-hand, examination of the topography involved when assessing the likely degree of integration or divorce between farming and herding.) One should also take into account the fact that, although many ancient settlements were, like Aigina, located in coastal areas a few hours' walk from the hills, many others were sited on medium-altitude inland acropolis sites far closer to regions of hill or mountain pasture. One well-known example is the village of Askra, situated on the E.S.E. slopes of the hill Pyrgani (Wallace 1974), only about 5 km. distant from the lower slopes directly below the peak nowadays called Zagoras, Hesiod's Mt. Helikon (Wallace ibid., 22), under which he was herding his lambs when visited by the Muses (*Theogony* 22-3). Like Koxios, Hesiod may have been utilizing grazing on the lower mountain slopes within easy reach of his fields in the Valley of the Muses.

In cases of movements outside the borders of one's *polis* the distances involved were also generally moderate. The inter-state agreements referred to above all concern neighbouring *poleis*, sometimes with small territories (Aigai and Olympos are only 9 km. distant: *RE* XVIII 1, 1942, 315). The fictional example in the *Oedipus rex*, although it involves two states whose centres were over 50 km. distant by land, also seems to presuppose rather shorter seasonal movements. Mt. Kithairon lies within 20 km. of the town of Thebes and there is reason to believe that the winter folds lay as near to the mountains as to the town, since at 760-2 the flocks are said to be pastured in the fields (*agroi*) far from the town. Similarly, Kithairon lies within 30 km. of the borders of the Corinthia. Contrast these distances with the 100 km. and more travelled by modern pastoralists in the N. Peloponnese (Koster 1976), the 250 or so km. travelled by Varro's Italian flocks (Garnsey 1988, 11) and the 800 km. covered by some sheep under the Spanish *Mesta* (Braudel 1975, 91).

Whether they took place within the territory of one *polis* or across those of several, few of the above vertical movements seem to have been designed to cross the boundaries of climatic or vegetational zones in order to exploit a markedly different ecological niche, as is the case in Alpine or Andean transhumance when the altitudes ascended sometimes attain thousands of metres, not just hundreds as in almost all the examples cited above. Despite the phenomenon of the compression of zones of vegetation relative to altitude on small mountains, such that the natural upper limit of woodland on southern Greek mountains is only about 1800-1900m. (Garnsey 1988, 20), comparatively few *poleis* contained mountains sufficiently high to benefit even from this factor or to create a significantly different climatic environment (cf. the useful Map 2 in Hammond 1981). Aigina, where Ag. Ilias is only 532 m. high, is a case in point. Moreover, as Garnsey points out (1988, 20-2), natural grassland is rare in Greece because of the hard limestone and marble rock which inhibit suitable soil formation. It is far from clear whether and to what extent man had by our period of antiquity engaged in deforestation in the mountains to create the artificial, pseudo-alpine meadows which exist in various locations today (cf. also Lewthwaite 1981, 60; Halstead 1987). In any case these pastures are frequently low in mineral and protein content and liable to summer drought, the standard feature usually ascribed to the lowlands. On all these grounds many of the vertical seasonal movements discussed above take on the aspect of an infield-outfield shift to uncultivable locations with unused grazing at a time of the year when the animals were not wanted near the growing crops (with the verticality of movement dictated by conditions of highly accidented topography) rather than a search for a different vegetational zone. They seem to have more in common with Delano Smith's 'reluctant transhumance' on the part of the arable farmers whose basic premise is that their animals should be fed from the local area as much as possible, and whose flocks and herds are therefore by preference sedentary until a move is imposed by local circumstances (1979, 231-9), than with the migrations of Koster's pastoralists-cum-agriculturalists who eagerly greet the high-level pastures of Mt. Ziria (Killini) in the northern Peloponnese as their real home (1976, 28). In short, many transhumant movements seem to have been determined more by economic reasons and by the nature of the local topography than by environmental or climatic constraints.

What kind of pastoral units did these movements involve? First, the bulk of the evidence relates to flocks of sheep or goats (cf. the examples in Robert 1949a, 154; *SEG* xi 377, with xxxi 328; and the Myania-Hypnia agreement), rather than to cattle, horses or pigs for which transhumance was usually either a less viable or a less necessary option. Secondly, and more importantly, the evidence appears to relate largely to flocks owned by wealthy men. This is clear both in the *Oedipus rex* and regarding the grants of *epinomia* to individuals. Jameson (forthcoming) suggests that it was substantial citizens who were pressing their interests in the second-century dispute over grazing land between Hermion and Epidauros and this may well have been the case in other border disputes. Robert (1949b) concludes that the herdsmen attempting to exploit the resources of the island of Herakleia in the third century were the agents of rich and powerful men.

The actual personnel who transhumed with these flocks were, it seems, normally either slaves or low-status hired freemen. This is the pattern in the *Oedipus rex* where the shepherds are a Theban slave and a Corinthian freeman characterized as a vagrant (1029). The common use of slaves as herdsmen is attested by a variety of sources – for example, by Euktemon's sale of the goatherd along with his goats (Isaios 6.33), by the seizure of the herders with the plaintiff's sheep in [Dem.] 47.52 (they are explicitly called slaves at 65) and by the frequent capture of slaves and livestock together in raiding (Ducrey 1968, 87ff.). Sokrates' remark that a vicious slave would not be entrusted with a man's livestock (Xen. *Mem.* 1.5.2) implies that a slave was the obvious choice as herdsman; what was at issue was the character of the slave to be chosen. Low-status freemen are also in evidence in [Lysias] 20.11, where Phrynichos is denigrated by being described as a youth who herded because of poverty in contrast to the respectable Polystratos; and in Dio Chrysostom (*Orations* 7.11ff.), where the two free herdsmen were landless men who had to take to mountain-farming and hunting when their few animals were seized along with those of their master. Consequently, the position of the free herdsmen mentioned in inscriptions, sometimes their own rough dedications, is uncertain. Some could be owners of their own flocks, but there is sufficient evidence for the employment of hired labour as an alternative to slave herdsmen to suggest that they could equally be servants of richer owners.

All this is precisely what one might expect in view of the suggestion above that it is only when livestock are kept on a sufficiently large scale that transhumance may become necessary, since it is often only then that resources on or near the farm itself become insufficient to support the animals in summer. In other words this evidence further supports the view that the primary factors governing the presence (or absence) of transhumance were not so much environmental as economic. There are also political grounds why one would not under normal circumstances expect many hoplite farmers to have migrated to summer pastures with their flocks. This was the period when their military services were often required on expedition or in defence of their arable lands. Nor is there any evidence for the small flocks owned by ordinary farmers being pooled together into communal transhumant herds. The intensive nature of agriculture discussed above obviated that necessity. The evidence accordingly points to the typical pattern being the movement of a particular marginal set of individuals – the herdsmen, whether slaves or hirelings, of wealthy men – not the mass of citizen farmers, still less the movement of entire herdowning populations as in the traditional seasonal migrations of groups like the Sarakatsani and Valtetsiotes in modern Greece. To what extent hired herdsmen migrated singly or with families is uncertain. The latter seems to be attested in the late first century A.D. in Dio Chrysostom's account of transhumance in S. Euboia; but the account in the *Oedipus rex* suggests that the hired Corinthian herdsman was alone on the mountain except for his fellow herdsman from Thebes. The remaining evidence is too vague to permit reliable conclusions.

The image of transhumance in ancient Greece presented by the above evidence, one of predominantly short-range infield-outfield movements of stock with comparatively little extensive pastoralism or long-range transhumance, is not difficult to explain. Long-distance transhumant pastoralism has usually been dependent upon specific historical circumstances, such as the weakness of lowland agriculture, a high level of market demand for pastoral products and a unified (and, ideally, *dirigiste*) political authority (Garnsey 1988, 19-20). These were the conditions which sustained the most large-scale transhumant systems of the western Mediterranean, the *Mesta* of Castille and the *Dogana* of Puglia. In modern Greece too the transhumant movements of the northern Peloponnesian Sarakatsani and the Arkadian Valtetsiotes into the southern Argolid depended upon the existence of large Turkish and monastic landholdings, the availability of pasture in marginal, malarial coastal zones and the presence of a rich market for pastoral products on the heavily-populated and comparatively non-agricultural nearby islands of Hydra, Spetsai and Poros, all under the overarching umbrella of Turkish, then independent Greek, rule. Both populations have been able to maintain their migrations in more recent times because, although citrus cultivation now occupies their former pastures, much marginal land previously used for subsistence cereal cultivation has been abandoned by a low density population whose farming is geared towards commercial production (Koster 1977, 76-80). Similarly, the large-scale transhumance of the Sarakatsani of N.W. Greece in the nineteenth century was dependent upon the depopulation and limited agricultural exploitation of the plains of Iannina, Arta and Preveza; since refugees from Turkey were resettled there as farmers in 1927, the Sarakatsani's situation has become increasingly difficult. More recently, their economy has relied upon the large Athenian urban market as an outlet for their pastoral products, sale of which permits the purchase of cheaper agricultural staples (Campbell 1964, 13-15, 248, 363-4).

Such conditions were largely absent for most of our period. First of all, Greek *polis* society rested predominantly upon a strong lowland agricultural base; and, since one of its prime characteristics was the incorporation of the peasantry, to a greater or lesser degree according to the nature of the political system, as citizens of self-governing communities, the lowland farming population was generally also in a position of some political strength. Moreover, although population densities must have varied considerably over both time and place, they were, especially in the Classical and Hellenistic periods, generally much higher than in most more recent periods, as is now being documented in a host of local studies (e.g. Hodkinson 1981, 270-9; Renfrew & Wagstaff 1982, esp. 145; Bintliff & Snodgrass 1985, 138-49, esp. 140-45; Sallares 1986, ch.2; Cherry, Davis and Mantzourani, forthcoming). The pressure of population upon limited lowland agricultural resources is attested by such major phenomena as emigration abroad and demands for redistribution of land which recur throughout Greek history. Secondly, in a world dominated by the ethic of household autarky and peopled largely by citizen-peasants engaged in mixed farming, one in which the infrastructure of trade beyond the local level was also relatively underdeveloped, the scale of demand for pastoral products must normally have had severe limits (cf. Finley 1973, esp. 107-110, 138) – sufficient at times to stimulate the small-scale, farm-based production discussed towards the end of Section I, but not the level implied by the practice of large-scale transhumance. Finally, the numerous small, autonomous polities of ancient Greece, each jealously guarding its own territory against unauthorized exploitation by outsiders, imposed a multitude of boundaries and obstacles in the way of long-distance movements. The very existence of, and necessity for, special bilateral agreements between individual *poleis* to regulate the movement of men and flocks, combined with the abundant evidence for hotly-contested border disputes, suggests that most animal movements must of necessity have been localized affairs within the boundaries of a *polis*; and, consequently, that transhumant herd sizes must normally have been restricted given this limitation on access to pasture.

All these reasons help explain the virtual absence from the Greek *poleis* of specialized pastoralist groups who in other periods have often maintained their large flocks by means of long-distance transhumance. Only under exceptional circumstances in the Archaic to Hellenistic periods were these various inhibiting factors absent – at Sparta (Section III below), although transhumance is not attested even there; perhaps also in Arkadia, noted for its epithet 'rich in sheep' (refs. in Hodkinson 1981, 280), where the juxtaposition of two strong *poleis*, Mantineia and Tegea, and numerous small weakly-organized highland communities who tended to come under the domination of their larger neighbours when Spartan influence was distracted, may at times have created opportunities for larger-scale transhumance for flocks owned by Mantineian and Tegean citizens (cf. Hodkinson 1981, 289, regarding Mantineia). Even these men, however, will have been landowners, not specialized pastoralists.

It is significant that the best documented case of transhumance comes after the end of our period, from the late first century A.D. in the territory of an unspecified city in southern Euboia (Dio Chrysostom, *Orations* 7.11-15). By that period lowland agriculture had declined to such an extent that it could be said that two-thirds of the territory was deserted and even the immediate suburbs of the town uncultivated; and it was proposed not only to allow both citizens and non-citizens to take up new land free of tax for ten and five years respectively, but to grant citizenship to any outsider who put 200 *plethra* of land (about 18 ha.) under cultivation (ibid. 34-8). The population had likewise severely declined and even the town had houses empty (34, 50). The regional social structure also seems to have been marked by greater inequalities of wealth than in pre-Roman times (cf. Finley 1973, 102). The owner of the transhumant herd had 'many herds of horses and cattle, many flocks, many good fields and many other possessions together with these hills' (11) around Cape Kaphereus near where his cattle were pastured in summer, a level of property surely rare in earlier periods and sufficient to attract the attention of the emperor, who apparently had him killed for his wealth. It was property-ownership of this magnitude which enabled him to engage in large-scale transhumance. Since he is described as a resident of the island rather than of the local *polis*, it is possible that he was an outside property-owner, such as is attested elsewhere in Greece under the Roman Empire (Kahrstedt 1954), as the significance of the political unity of the Empire increased at the expense of local autonomy. Needless to say, the combination of circumstances which permitted this pattern of husbandry contrasts sharply with those prevalent in our period.

III *Animal husbandry, subsistence and surplus*

What implications does the above evidence have for the question of the significance of animal husbandry in the procurement of subsistence and surplus? An attempt at a complete answer to this question lies beyond the scope of this paper and would in any case be hampered by the paucity of the literary and epigraphical evidence and the absence of usable zoo-archaeological studies of animal bone assemblages from mainland and Aegean Greece during our period (Payne 1985 & pers. comm.). The question, nevertheless, demands some considered response, even if the answers must necessarily remain tentative and provisional.

a) *Animals and subsistence*

The issues raised can be approached from two contrasting but complementary perspectives. On the one hand, there is the general significance of animal husbandry within the Greek economy as a whole. Sallares (n.d.) has suggested that the role of animal husbandry in ancient Greece, and particularly in Attica, can be judged to be relatively insignificant within a comparative historical context, especially when set beside its role in the agricultural system of early modern Britain, because of the absence of ley-arable rotation, the comparative shortage of meadow land and fodder crops, and also because the ancient Greek animal population density will have been inversely correlated with the high density of human population owing to competition between men and animals for comparatively scarcer territorial resources. Garnsey (1986, 10) has observed that even on occasions in Mediterranean history when pastoralism has established a position of relative independence in relation to the agricultural economy, the mass of population has continued to depend upon agriculture for its subsistence. Halstead (1981, 314; 1984, 313-5) has calculated that, even when exploited for their secondary products as well as meat, a single individual relying solely upon pastoral products for his subsistence would have to run a flock of at least 25 sheep, whereas he might support himself off only 1 ha. of arable land. Foxhall and Forbes (1982, 74) have estimated that in classical antiquity grain alone probably contributed about 70-75% of the calories of the average diet. Moreover, evidence has already been cited which suggests that in some *poleis* at least many estates may not have possessed livestock other than draught animals. On these various grounds it would appear that the overall economic significance of animal husbandry was not particularly high.

Within that general framework, however, important questions remain concerning the significance of animal husbandry for those households which did own stock. From this perspective issues concerning the absolute numbers of animals within the whole economy or the value of pastoral products as a proportion of overall agricultural calorie production or GNP may be less important than those concerning the function and significance of animal husbandry within particular household economies. Here it is necessary to distinguish various types of function and degrees of significance, since the foregoing discussion suggests the existence of considerable variation in types and scales of animal husbandry, variations related largely to differences in levels of landownership.

These variations can perhaps best be understood against the background of a factor critically evident to any owner of livestock, namely the inherent riskiness of pastoral enterprises compared with other agricultural options. Koster has highlighted the fact that before the advent of modern concentrated feeds and veterinary drugs in Greece after World War II an unusually dry autumn or outbreak of disease could quickly decimate a flock (1977, 163, 241-2, 316-9). The high risks of animal husbandry are relevant to the question of which animals are reared. Here we may draw upon some findings from the comparative ethnographic study of pastoralism by Dahl and Hjort (1976). They note that cattle are less prone than ovicaprines to disease and drought; but because they breed and mature at a slower rate it takes longer to reestablish a decimated herd. Ovicaprines are more vulnerable but have a considerably faster rate of regeneration and growth (28-75, 87-112, esp. 111). They observe that the preferred strategy of pastoralists in many parts of the developing world is the exploitation of several complementary species of livestock, both small and large, with ovicaprines typically providing meat and cattle milk. Poor households which have lost all or many of their animals often invest initially in ovicaprines because of their higher growth rates, aiming subsequently to introduce larger livestock for longer-term stability and security (232-4).

It is illuminating to compare this pattern drawn from pastoral-based societies with those prevalent in the arable-based household economies of ancient Greece. What little evidence there is for ordinary citizen households suggests the maintenance of very modest numbers of small livestock, the scale of the enterprise being limited partly by the demands of husbandry upon family labour and partly by the limited agricultural residues and grazing available on small estates. Whether such households also maintained larger livestock would normally depend upon the possession of sufficient land to justify the employment of draught oxen. The value to peasant families of such small-scale stockkeeping as a multi-purpose nutritional resource, as a provider of manure, as a means of transforming otherwise indigestible plant material (whether wild or the residue from cultivation) for human use and as an emergency store of food in time of famine is well known to students of traditional farming (e.g. Koster 1977, 364; Lewthwaite 1981, 62; Barker 1985, 42). Here we can detect some significant differences from pastoral-based household economies, with ovicaprines supplying a wider range of supplementary subsistence services orientated more towards the production of milk products and the maintenance of arable production; and with cattle largely confined to the latter role.

It is possible that all these factors gave animal husbandry a more significant supplementary role in the subsistence economy of the average citizen household than would appear from a simple calculation of the proportion of calorific intake provided by animal products. One important advantage in a situation in which ovicaprines were not depended upon annually as the major source of subsistence or surplus will have been the high level of flexibility which they afforded. Under what might be termed the "concertina model" of Mediterranean animal husbandry, the stock could be exploited flexibly each year, as varying climatic conditions affected the amount and quality of available grazing and fodder, being kept on bare subsistence in bad years, exploited more fully in good ones. Strategies could also be adapted as biological development altered the size and composition of the household and with it the amount of available labour and the nature of its nutritional needs. In households in which the balance between people and resources was delicate, the direct and indirect contribution of a few small livestock may often have made a critical difference in enabling a family to achieve either its regular subsistence; or, in the case of somewhat better-off households, to create a small surplus sufficient for the maintenance of hoplite status. The evidence from Keos suggests that under some circumstances even small numbers of sheep could bring not insignificant profits from the sale of cheese and lambs. Finally, in times of crisis livestock might be a crucial reserve. For example, Xenophon's *Hellenika* (4.5.1) states that during the military engagements of the Corinthian War, which deprived those in the town of Corinth of the use of much of their land, many of the citizens were fed from their livestock which were being pastured in Peiraion. (His account in the *Agesilaos* (2.18) is more nuanced, emphasizing also the Corinthians' use of the area's crops.)

Rather further down the social scale, on the two-family marginal upland farm attested by Dio Chrysostom, we can detect a different strategy of animal husbandry. This poverty-stricken group appears to have relied as heavily upon hunting and animal husbandry as upon their limited arable crops.[13] In keeping with their paramount need to ensure their subsistence, their stock of eight she-goats was complemented with the extra security of a hornless cow with calf. To judge from the absence of male goats, their strategy (Payne 1973) seems to have been aimed at optimizing milk production for home consumption, in combination with the slaughter of male kids for meat (note the small goatskins mentioned at 43) and perhaps sale outside the household for the procurement of other necessities (cf. the purchase of wine referred to at 76). Note also the absence of sheep, a pattern of stockholding common in recent centuries among poor households on Corsica and Sardinia, where the sheep was invariably the first animal to be jettisoned, the cow the last (Lewthwaite, unpub. MS). Goats yield more milk than sheep over a longer period, reproduce at a faster rate (Dahl & Hjort 1976, 112, 210-12) and were no doubt better adapted to the upland terrain and less likely to sicken in its cold winter environment. As was observed above, the cow is the most secure animal asset. The combination of the two was clearly aimed at minimizing the risks of starvation.

b) *Livestock upon wealthy estates*

The patterns of husbandry observable on the estates of wealthy landowners are, not surprisingly, different again. Here sheep had a more considerable role. The relative proportions of different animals on larger estates must have varied considerably according to the potential grazing available. In areas of lower population density and more abundant high quality grazing a more varied combination was perhaps more likely. For example, the 20,000 or so livestock of every kind captured at the fortress of Pyrgos in the comparatively lush region of Elis by Philip V of Macedon in 209 (Livy 27.32.9) is likely to have included at least some wealthy men's herds, despite the large number of people, 4000, accompanying them. But, as so often in Greek history, the largest body of precise evidence comes from Attica, where the limited availability of wetland grazing severely reduced the numbers of cattle compared with sheep and goats, even on larger estates. Reference has already been made to the flock of 50 sheep owned by the plaintiff of [Dem.] 47. The property of Theophon, who left 2 talents' worth of land, included 60 sheep and 100 goats (Isaios 11.41). The confiscated property of one of the Hermokopidai included 84 sheep and 67 goats, both with young in addition, but only two work oxen and six further cattle plus calves (Pritchett 1953, 272: VI 68-73; 1956, 255-60). Even in Attica, however, there is some evidence for an alternative pattern. Nausikydes, the slave-owning miller, is said to have invested his profits in large herds of pigs and cattle (Xen. *Mem.* 2.7.6); and another of the properties of the Hermokopidai contained a pig sty and cow barn (Pritchett 1953, 265: V 39-40).

A notable feature of the above figures is the relatively modest herd sizes. The universally small flocks of sheep of Kean farmers have already been noted and comparable sizes are suggested by Hellenistic grants of *epinomia*. The late third-century grant by the *polis* of Akraiphia to one of its citizens to whom it was in debt made provision for the pasturage of only 50 livestock (*SEG* xxix 439). In second-century Kopai two female creditors were given free pasturage for 200 livestock each (*SEG* xxii 432). Only in the grant of third-century Orchomenos to a certain Euboulos from the nearby state of Elatea do we find mention of significantly larger herds, 220 large livestock and 1000 small (*SEG* xxvii 63). As a creditor of another state, however, Euboulos may have been a man of considerable regional importance and therefore not necessarily typical. (The inscription, unfortunately, does not make his status clear.) That the pattern of modest-sized herds was a more widespread phenomenon is suggested by Richter (*RE* X.A, 1971, *s.v.* 'Ziege', 408), who points out that ancient Greek had only a single word, *aipolos*, for the herder of small livestock, a fact which he ascribes partly to the small size of mixed herds.

This evidence stands in marked contrast to the much larger ovicaprine herds attested in parts of modern Greece before the advent of concentrated feeds and modern veterinary medicine significantly lowered the risks of pastoralism (Koster 1977, 163-8). This should serve as a further reminder that traditional extensive methods of Mediterranean animal husbandry are not necessarily a reliable guide to practices in our period which seem often to have been rather more intensive. These much larger herds belonged to independent subsistence shepherds mainly without alternative resources who needed the insurance of large flocks to maximize their security. The capability to maintain them depended upon the availability of a sizeable labour force, provided either by a large number of sons (cf. the case-histories in Koster 1977, 221, 452, 466) or in the case of the Sarakatsani by a social institution, the *tselingato*, a patrilineal clan association which emerged (and dissolved) according to specific labour needs (Campbell 1964, 8, 16-18; Koster 1977, 77-8). This in turn depended upon the utilization of large blocks of winter grazing in a period of low population density. As grazing became available in smaller units in N.W. Greece from the 1930s and in the southern Argolid in the late 1960s and early 1970s, so the size and use of multi-household associations declined.

This analysis of traditional modern pastoralism highlights the conclusions of recent research, first, that the crucial factors in determining flock size are the amount and quality of land and, secondly, that there is generally more concern for the ratio of herders to animals than for a fixed optimum number of stock (Dyson-Hudson 1980, 27). It suggests by contrast a number of reasons for the limited size of rich citizens' herds in antiquity. First there is the heavy reliance upon grazing resources situated in densely populated agricultural regions where the holdings were modest in overall extent and often fragmented into smaller plots. In Attica, for example, the fact that even wealthy men normally owned a maximum of about 300 *plethra*, approximately 27 hectares (Burford Cooper 1977/78, 170), must have put a ceiling upon the amount of grazing and fodder available from a man's private estate, especially given the heavy feeding demands of working oxen. Furthermore, the evidence for states granting the privilege of *epinomia* to their own citizens implies that normally citizens had to pay fees to graze their animals on public pasture (cf. Marek 1984, 149). This may have proved a significant disincentive to the maintenance of larger herds. Nor could even larger landowners always afford to take out a long lease on temple or public land. Although grazing may have been permitted upon fallow and stubble on other men's land, as on modern Methana (Forbes 1982), that itself would have necessitated small flocks and close supervision to prevent damage to boundaries and neighbouring properties. Koster observes that in the modern Argolid one active shepherd can effectively herd 50-60 sheep on agricultural land (1977, 255), a figure closely similar to flock sizes in classical Attica. This points to a second factor, namely the predominant pattern of single herdsmen rather than larger groups. This itself may in part be the effect of the practice of grazing on agricultural land. But since many herdsmen were slaves, it may also reflect a fear of groups of unchained slaves moving around the countryside.

Thirdly, the keeping of animals for strictly economic purposes was only one aspect of animal husbandry on wealthy estates. Many wealthy men were also involved in the expensive maintenance of horses in order to serve in the cavalry. Theophon, whose case has already been cited, is a good example, since in addition to sheep and goats he maintained a fine horse which he rode when he was a cavalry commander (Isaios 11.41). Some of the very wealthiest men even bred horses to participate in elite equestrian sporting events, an activity which became particularly important with the dramatic increase in the number of chariot-racing contests in Panhellenic games during the first third of the sixth century (Davies 1981, 103ff.; on the expenses of horse-breeding, Davies 1971, xxv n.7; cf. also the list of Olympic victors in equestrian events in Moretti 1957 & 1970). The horse was the useless animal *par excellence* in an era before the invention of the harness enabled it to be used for traction. Hence raising horses was a means of expressing social distance and became a characteristic attribute of the rich (Sallares 1986, ch. 3, iv). Aristotle also states (*Pol.* 1289b33-5) that the extent of one's horse-ownership was a mark of internal differentiation among the wealthy.

These considerations indicate some of the limits upon the significance of animal husbandry in the procurement of surplus and profit by wealthy households. They do not, however, eliminate the phenomenon. The frequent evidence for flocks of soft-fleeced sheep, obviously reared for the sale of high-quality wool, is sufficient testimony in this regard. Wealthy arable farmers are unlikely to have kept such flocks had it not been profitable to do so. Indeed, it could be argued that the very prominence of modest-sized herds of rapidly reproducing, but comparatively vulnerable ovicaprines, often intensively stall-fed, signifies a high degree of profit motive – an emphasis upon maximizing the return per head in good years, rather than aiming at a lower, steadier rate of return. Koster's work (1977, 245, 266ff.) has demonstrated the potential profitability of intensive pastoralism of comparable sizes of flocks (mean size 50-70 sheep) to those attested in our period. Although ancient landowners did not have the advantages of modern concentrated feeds to increase milk production, the semi-intensive methods of husbandry discussed above may have produced sufficiently worthwhile profits sufficiently often to make a not insignificant contribution to the income of wealthy households.

These suggestions find some support in the ancient evidence. If the Kean farmer with his small flocks could achieve surplus and profit, then so obviously could owners of somewhat larger herds. A character in Antiphanes' play *The seamstress* (Edmonds 1957-61, II 20, *apud* Athen. 402b-e) expresses his preference for eating a sheep which has neither wool nor cheese and a goat which produces no cheese, on the grounds that 'the profits derived from full-grown animals are such that I can put up with eating these poor ones'. In Xenophon's *Memorabilia* (4.3.10) Sokrates claims, somewhat exaggeratedly, that men gain food and produce wealth (*chrēmatidzontai*) not less from animals than from crops. In mid-fourth-century Epidauros a certain Timainetos dedicated a small altar to the Anakes as 'a tithe of the goats', probably of the profits from a herd of which he was the owner (*SEG* xxvi 451; Jameson, forthcoming). References in Athenaios'

Deipnosophistai to the export of cheeses from such places as Sicily and Tromilia in Achaia (27d & f, citing Antiphanes and Hermippos; 658a-c, citing Philemon, Demetrios of Skepsis and Semonides of Amorgos) and to the use of cheese as an item in expensive dinners (146e & 364d = Menander, *The carouse*, Edmonds 1957-61, III.B 319) testify to profitable animal husbandry on the part of wealthy men. So does the abundant evidence for the exploitation of wool. Although some of it was to be put into the household's store (Xen. *Mem.* 2.9.4), to be used later no doubt for producing the household's clothing (Xen. *Oik.* 7.6, 21, 36), some could be sold to households without their own supply (Xen. *Mem.* 2.7.12). Moreover, prosperity from sheep-rearing must have gone hand-in-hand with the production of the famed Megarian one-sleeved woollen tunics (Ar., *Acharnians* 519; Xen. *Mem.* 2.7.6; Legon 1981, 87ff.) and the highly-regarded Milesian wool (Michell 1957, 69). The herds of pigs and cattle in which Nausikydes invested his profits from milling are themselves likely to have been profit-producing, since the animals upon which he chose to concentrate were probably reared only in moderate numbers in Attica, as Jameson's essay in this volume indicates. Jameson also notes the Megarians' customary sale of pigs to the Athenians of which Aristophanes made comic use (*Acharnians* 729ff.). His essay, moreover, which explains how the timing of periods of peak demand for sacrificial victims was linked to the optimum seasons for the disposal of animals from private herds, demonstrates the significance of the sale of all kinds of domesticated livestock as an extra source of income from animal husbandry.

The above evidence indicates that ancient Greek landowners were alert to the advantages of livestock as a form of wealth which, compared with land, contained a more fluid potential for increase, disposal for profit and subsequent regeneration and which, especially in the case of sheep, could be exploited flexibly for milk, fleece, meat or direct exchange (Halstead 1984, 351), according to variations in economic conditions. The relative importance of the pastoral element as a producer of profits on the combined agro-pastoral estates of wealthy landowners is of course impossible to quantify. But the evidence suggests a similar situation to that recently hypothesized by Halstead (1987) concerning the returns from cereal agriculture. He has suggested that precisely because of the variability and risks inherent in agriculture (factors present to an even greater degree in animal husbandry), although returns from cereals were normally low, occasional windfall profits could be expected by some farmers in years of famine. Although pastoral products were not staples as were cereals, a milder version of this phenomenon is not unlikely in times of heavy animal losses due to drought or disease. Note, for example, the escalated prices of sacrificial sheep, goats and cattle in the Eleusinian inscription of 329/8 during the famine of 330-26 (*IG* II^2 1672.289-90; Pritchett 1956, 256, 258, 260). The scale of profit thus periodically generated may have made animal husbandry an important supplementary source of revenue, thereby explaining why many arable landowners did maintain animal herds. The extremely provisional nature of these remarks should, however, be emphasized. Before more secure conclusions can be advanced we need to know considerably more than we presently do about the level and nature of local, regional and Mediterranean-wide demand for surplus pastoral products and about the mechanisms of exchange for the disposal of such products in all stages of transfer from herdowner to consumer.

The constraints on size considered above did not apply at every time and place during our period. The case of classical Sparta provides an instructive series of parallels and contrasts to the picture delineated above. Owing to their possession of the large territory of Messenia and a system of land tenure and inheritance which led to an increasing concentration of land (Arist. *Pol.* 1270a15-b6; cf. Hodkinson 1986), leading Spartiates owned private estates considerably larger than those of their counterparts in most other Greek *poleis* upon which they were able to raise much larger herds, as [Plato] *Alkibiades* I makes clear,

> They have land both in their own country [i.e. Lakonia] and in Messene. Not one of our estates could compete with theirs in extent and excellence, nor again in ownership of slaves, both helots and others, nor yet of horses, nor of all the other livestock that graze in Messene (122d).

The Messenian lands upon which these livestock grazed were

> watered with countless streams, furnished with good pasture for both cattle and sheep, being neither very wintry in the blasts of winter, nor yet made too hot by the chariots of Helios (Euripides, *apud* Strabo 8.5.6).

But it was not only favourable material and environmental circumstances which prompted leading Spartiates to raise livestock on a grand scale; nor was it primarily the same kind of profit motive discussed above. Animal husbandry was locked into the sphere of liturgies and largesse to an extraordinary degree. In the common meals rich citizens were able to impress their poorer messmates by making extra donations of meat drawn from their flocks (Athen. 139c, 140c-e, 141e, citing a variety of sources; cf. Hodkinson 1983, 251ff.). They could also gain prestige through the system of institutionalized sharing, according to which poorer Spartiates were permitted to borrow their horses and hunting-dogs when in need of their services (Xen. *Lak. Pol.* 6.3; Arist. *Pol.* 1263a35-6). After the introduction of a cavalry force in the late fifth century, it was again the very rich who provided the horses for others to ride (Xen. *Hell.* 6.4.11). It is no surprise, therefore, that Agesilaos II, a king with a consummate ability for gift-giving in return for influence, stocked his estate with many hounds and war-horses (Xen. *Ages.* 9.1); and evidently oxen too, since he invariably gave an ox as a gift to new members of the *gerousia*, many of whose votes he could control at critical moments (Plut. *Ages.* 4.5; Xen. *Hell.* 5.4.25-33). Finally, in a society in which many of the normal aristocratic outlets for expenditure and display were denied to them, it was through chariot-racing that leading Spartiates were able to demonstrate their wealth and standing. The unmatched string of Spartiate successes at Olympia, especially in the late fifth century (Moretti 1957 & 1970; Ste. Croix 1972, 354f.), testifies to the scale of this activity in the Panhellenic sphere. Damonon's inscription (*IG* V 1, 213), set up in Sparta's central sanctuary of Athena Chalkioikos, which records an impressive series of victories in contests inside Lakonia and Messenia and which boasts proudly that they were achieved with his own fillies bred from his own mares, is sufficient evidence that chariot-racing and its associated horse-breeding was intended to impress an internal as well as an external audience.

IV *Further research*

The case of classical Sparta, divergent though it is in several respects from the overall patterns of animal husbandry discussed earlier, suggests a number of insights to be applied in further research. The first and foremost lesson is that the study of animal husbandry needs to be pursued on a broad front in relation to a range of political, socio-economic, religious and military factors. The connections between the character of a *polis* and the nature of the animal husbandry practised by its citizens may be less immediately obvious in other more pluralistic *poleis* than in a highly organized, relatively uniform society like Sparta; but their study nevertheless remains essential. A comprehensive view is necessary in order to understand fully the role and significance of animal husbandry.

One implication of this viewpoint is that a more sophisticated and thorough-going study of ancient Greek animal husbandry would reveal many variations and nuances, both from *polis* to *polis* and from period to period, not encompassed by the highly generalized perspective necessarily adopted in this preliminary investigation. Detailed analysis of the various categories of literary and epigraphic evidence briefly referred to above, several of which remain largely unexploited, would no doubt illuminate specific aspects of these regional and temporal variations. As always, however, the difficulty which faces the ancient historian is the extreme paucity of continuous literary and epigraphic testimony for the vast majority of Greek *poleis*. This situation calls of course for a widening of the range of evidence to be exploited, something which is only just beginning for our period. The absence of references to faunal evidence in this paper, for example, reflects not the historian's blindness to non-documentary evidence but the failure of classical archaeologists, with recent notable exceptions, to engage in the systematic collection of animal remains on ancient Greek sites (Payne 1985, 211). The systematic exploitation of this kind of evidence is vitally important not only in order to shed light upon the role of pastoralism in those *poleis* poorly covered by the documentary sources; but also, by means of comparative studies, to reveal the general range of regional and chronological similarities and variations, which will then lead to a better understanding of those societies about which we are already partially informed. The equally pressing need to pay due attention to plant remains, whose relevance to our subject should be evident from the earlier parts of this paper, has been given fresh urgency by Sallares' remarkable recent study (1986, ch.3, vi-xii) of the socio-economic impact of the biological evolution of cereal plants in antiquity. The implication of the need to study animal husbandry within the overall context of Greek society calls indeed for the widest inter-disciplinary collaboration of historians, archaeologists, botanists, anthropologists, geographers and many others. Only through the use of the broadest range of evidence will the comprehensive view advocated above ever lie within our grasp.

Given the necessarily long-term process of integrating the research of diverse disciplines, it is appropriate for historians to consider the contribution that can meanwhile be made by practitioners of the more conventional and well-established approaches to the study of ancient history. I should like to suggest that one potentially fruitful method of illuminating the chronological and regional variations in the character of animal husbandry could be by relating them to their political context, a subject about which we often have more evidence than about other aspects of ancient society. It is in the political sphere that the issue of chronological and regional variation is most starkly posed, owing to the sheer number of changes of political system experienced by each of several hundred *poleis* from the archaic period down to their incorporation within the Roman empire. In such small-scale societies, in which all citizens were potentially both members of the ruling body and owners of land and livestock, political changes were often likely to have had an impact upon the practice of animal husbandry.

These observations underline my final and more general point that we need to move away from the stereotypes and simple models of ancient Greek animal husbandry which have tended to prevail in the past. In reacting, as this essay does, against the old stereotype of a near-universal divorce between animal husbandry and agriculture, it is not my intention to substitute a new simple model of the relationship between pastoralism and Greek society. It *is* my intention to stress that we need to be specific about the complex and changing political, socio-economic and environmental contexts of animal husbandry and sensitive to the varied options and responses open to owners of livestock within those contexts.

UNIVERSITY OF MANCHESTER STEPHEN HODKINSON

Acknowledgements: I should like to thank the following friends and colleagues without whose generous contribution of ideas, references or permission to cite unpublished material this paper could not have been written: Lin Foxhall, Peter Garnsey, Paul Halstead, Hilary Hodkinson, Alistar Jackson, Michael Jameson, James Lewthwaite, Christian Marek, Sebastian Payne, W.K. Pritchett, Bjørn Qviller, Peter Rhodes, Tracey Rihll, Robert Sallares, John Salmon, Tjeerd van Andel and David Whitehead. I am also glad to acknowledge my debt to the discussion at the Bern congress, especially the comments of Jens Erik Skydsgaard whose own essay in this volume will present a rather different view of several of the issues which we discuss. Although my paper has been written without knowledge of the final details of Dr Skydsgaard's essay, it is in part a response to the critical and valuable stimulus of our discussions in Bern.

NOTES

1. One exception appears to be the slave-owning miller, Nausikydes (Xen. *Mem.* 2.7.6), who invested his business profits in large herds of pigs and cattle. This is hardly a case of specialized pastoralism, especially as he may also have invested in the land on which to keep his herds (Davies 1971, 315); but it does serve to remind us of the variety of circumstances in which animal husbandry might be practised.

2. I labour these points in order to avoid the mistaken impression (which surfaced at the Bern congress) that, disagreeing as I do with the view which takes the first model as a universal representation of the realities of ancient Greek agriculture, my own analysis must therefore correspond exactly with the features described in the alternative model. My use of the models, and of the second model in particular, is as a heuristic device for structuring my discussion and for directing attention to neglected aspects of agriculture and pastoralism, not as an *a priori* statement of opinion. A model is 'a simplified structuring of reality which presents supposedly significant relationships in a generalized form... This selectivity means that models have varying degrees of probability and a limited range of conditions over which they apply.' (Chorley & Haggett, as cited by Finley 1985, 60). This quotation encapsulates my attitude to the relationship between the two models and the realities of ancient Greek agriculture and animal husbandry.

3. Although Theophrastos' works *Enquiry into plants* and *On the origins of plants* are most obviously classified as botanical treatises, they can also be used as evidence for agricultural practices. Although in Varro's *Res rusticae* (1.5.2) Stolo is made to say that Theophrastos' books 'are not so well adapted to those who wish to tend land as to those who wish to attend the schools of philosophers', he hastily adds that this 'is not to say that they do not contain matter which is both useful and of general interest (*et utilia et communia*)'. The Theophrastos' works are not manuals of *guidance for* farming does not remove their value as a source of *information about* farming. Not only did he draw upon information in treatises on farming such as the *Geōrgika* of Androtion and Kleidemos, in addition to the unnamed works of contemporaries (*H.P.* 2.7.2-3; 3.1.4; *C.P.* 1.10.3; 3.10.4; 3.23.1; 5.9.10 = FGrH 324F81-2; 323F33-6, with the comments of Jacoby 1954, I 108 and 85; Einarson & Link 1976, xix-xx; cf. Rostovtzeff 1941, II 1182f., who characterizes Theophrastos' works as 'collections of the maxims of practical farmers, systematically classified and logically interpreted'); but, equally importantly, his basic perspective 'does not consider a plant in isolation, but asks what its relationship is as a living

organism to sunshine and exposure, soil and climate, water and cultivation', with 'a particular interest ... in the response of the plants to domestication' (Hughes 1975, 122, 123; cf. Einarson & Link 1976, ix-xxiii). Some parts of Theophrastos' works are, indeed, more agricultural than botanical, as (for example) the sections on the propagation and cultivation of trees (*H.P.* 2.6.12-8.4), on the sowing and manuring of field crops (*H.P.* 8.6.1-3) and, in particular, the long section on the aims and procedures of agriculture (*C.P.* 3.2.1-4.16.4). On the 'angewandte Botanik' which dominates Theophrastos' approach, see most recently Wöhrle 1985, 33-4 & esp. 46-51.

4. Cf. *LSJ*[9] and Chantraine 1968-80, *s.v. aphakē* (tare, *Vicia sativa*, var. *angustifolia*); *erebinthos* (chickpea, *Cicer arietinum*); *thermos* (lupin, *Lupinus alba*); *kuamos* (broad bean or horsebean, *Vicia faba*); *lathyros* (chickling or grass pea, *Lathyrus sativus*); *orobos* (bitter vetch, *Ervum ervilia*); *pisos* (pea, *Pisum sativum*); *phakos* (lentil, *Ervum lens*); *ōchros* (*Lathyrus ochrus*). The identifications of the ancient Greek names with the English and Latin botanical names are those given in Hort 1916, II 437-99, not all of which should be regarded as certain.

5. Books II and VI of the *Epidemics*, which both report this episode, are best dated to the last decade of the fifth century or first half of the fourth century (Grmek 1983, 448-52).

6. To the literary references to bitter vetch, add the entry in one of the inscriptions recording the sale of the personal property of the men convicted of mutilating the Herms and profaning the Eleusinian mysteries in 415/4 B.C. The crops from one estate at Eretria included 7 *phormoi* of bitter vetch along with one *phormos* of broad beans, 15 *phormoi* of wheat and a *medimnos* of barley (Pritchett 1953, 251, II 90-5). According to Pritchett's estimates (1956, 199), the two pulse crops accounted for more than one-third of the value of the crops.

7. Although some modern research (Arnon 1972, I 465) suggests that harvesting a seed crop from legumes has deleterious effects upon the yield of a following grain crop (a drop of at least 30% in yield as compared with grain following fallow) because the legume removes more nitrogen than the root nodules produce, it is uncertain how much this was appreciated in Greek antiquity. Theophrastos (*H.P.* 8.7.2; 9.1) states that broad beans reinvigorate the soil even when they are sown thickly and produce much grain. Elsewhere (*C.P.* 4.8.1-3) he generalizes that, apart from the chickpea, legumes do not take strength from the cultivated soil but rather give it new strength.

8. Oder (*RE s.v.*, I 1064-5) suggests that Aischylides perhaps lived in the third century, but unfortunately there is no certain evidence for this; nor is it known from which of the *poleis* on Keos he originated.

9. Lin Foxhall has pointed out to me an uncertainty concerning Aischylides' meaning here. Olive trees are not deciduous and, although the leaves do eventually fall, they are not very edible at that stage. Aischylides may be referring to leaves brought down by the beating of the trees which was part of the process of harvesting the olives, or possibly to prunings.

10. Neither source gives the precise time of year. Walbank (1936, 69, 71) has dated the episode at Kynaitha to spring (more precisely, May) 241. The episode in [Dem.] 47 is more problematic. Some generic indication is given by the fact that the plaintiff's sheep were seized a few days after he had as

trierarch despatched the trireme urgently required for the personal use of the *stratēgos* Alkimachos (49-52). This seems to indicate that a naval expedition was already in progress, since the verb *apostellō* used at both 50 and 51 surely means that the trireme was to be despatched from Attica to wherever Alkimachos already was. Hence the time of year should be the campaigning season, that is, spring, summer or early autumn. Greater precision is impossible, partly because nothing more is known about Alkimachos' expedition, partly because of uncertainty concerning the time of year at which the new trierarchic year began. (I am grateful to Professor P. J. Rhodes for advice on these points.)

11. My equivocation over the identity of the urban centre reflects disagreement concerning the location of the Hippodrome. It has commonly been located in the region of modern Nea Phaleron near Peiraieus (e.g. Judeich 1931, 456); Benton (1972), however, locates it just east of the hill of Kolonos Hippios not far from Athens itself (cf. Thuc. 8.67.2). The precise point of departure of Theophemos and his associates on their thieving expedition, which would help to solve the problem, is itself uncertain. It may have been Peiraieus rather than Athens, since it was there that the plaintiff heard the news of the thefts (62), having met Theophemos in person an unknown length of time earlier that day (51). Cf. also 26, where the two men met near the Hermes by the postern, probably near the City gate in the north fortification wall of Peiraieus (Judeich 1931, 152-3). On at least one occasion (62), however, they did meet in Athens.

12. On the problems involved in determining the time of year of this incident, Pritchett 1969, 63 n.120, who inclines towards a date in July (ibid. 44). This is reconcilable with what we know of the Athenian calendar, should Plutarch's statement that 'the conflict at Mantineia has made the twelfth of Skirophorion more sacred' be taken as specifying the date of the battle rather than of a commemorative ceremony. The published archon tables of Dinsmoor, Meritt and Meritt-Traill (refs. in Woodhead 1981, 140-1 nn.17-18) allow Hekatombaion 1 to fall on the new moon before the summer solstice or the first or second afterwards (Pritchett 1970, 47-8, 64-6). According to the tables in Parker & Dubberstein (1956), new moons in summer 362 would fall on 26 May, 24 June and 24 July. If 1 Hekatombaion 362/1 fell on 24 July, 12 Skirophorion 363/*2* would be approximately 6 July; if the former fell on 24 June, the latter would be around 6 June. (I am grateful to Professor W.K. Pritchett for advice on the above points.)

13. This is suggested, in particular, by consideration of the ratio of the number of family members to available foodstuffs. It is possible to detect at least three stages in the development of the two-family unit. 1) At the time when the two herdsmen were deprived of their occupation it is said (15-16) that a small plot of land near the summer steading was sufficient for their families, which then probably comprised 4 adults and up to 4 unmarried children (20). 2) Later, at the time the speaker, son of one of the ex-herdsmen, was summoned to town, the families comprised at least 7, possibly 8, adults plus possibly at least one child (58, 68). (The two ex-herdsmen and at least one of their wives were still alive: cf. 47, where the 4 sickles and mattocks correspond to 4 adult males; also 20, with 64, which indicates that the episode in the town must have taken place before the ex-herdsmen's deaths, which occurred about a year before Dio's visit, since the vegetable and tree garden which Dio saw had then not yet been planted.) Yet their total quantity of foodstuffs in store was merely the inadequate amount of 2 *medimnoi* of wheat, 4 of barley, 4 of millet and 1/3 *medimnos* of broad beans (45; on consumption requirements for grain see Foxhall & Forbes 1982). This is the stage to which the evidence for the families' livestock (discussed in the main text) relates. 3) Later still, at the time of Dio's visit, there were 5 adults, a boy and girl of marriageable age, and at least 4 younger boys (20, 65-75). Although it was claimed that they needed nothing, they had had to borrow the seed for their wheat crop the previous year (69). On the important role of hunting in the provision of subsistence, cf.

esp. 44, with 3-5, 16-20, 43, 50, 57, 65, 67, 71-73; on animal husbandry 47, 72. (All the references above and in the main text are to *Orations* 7.)

Bibliography

Arnon, I., 1972. *Crop production in dry regions.*

Barker, G., 1985. *Prehistoric farming in Europe.*

Bell, M. & Limbrey, S. (edd.), 1982. *Archaeological aspects of woodland ecology*, *BAR* Int. Ser. 146.

Benton, S., 1972. 'Echelos' hippodrome', *BSA* 67, 13-19.

Bintliff, J.L., 1977. *Natural environment and human settlement in prehistoric Greece*, *BAR* Int. Ser. 28.

Bintliff, J.L. & Snodgrass, A.M., 1985. 'The Cambridge/Bradford Boeotian expedition: the first four years', *Journal of Field Archaeology* 12, 123-61.

Bolkestein, H., 1958. *Economic life in Greece's golden age*, New ed.

Braudel, F., 1975. *The Mediterranean and the Mediterranean world in the age of Philip II*, 2nd ed.

Brendel, O., 1934. *Die Schafzucht im alten Griechenland*, Diss. Giessen.

Brouk, B., 1975. *Plants consumed by man.*

Buck, C.D., 1955. *The Greek dialects.*

Buckler, W.H., 1934. 'Une borne Ionienne', *Revue Philologique* 3rd ser., 8, 293-8.

Burford Cooper, A., 1977/78. 'The family farm in Greece', *CJ* 73, 162-75.

Chantraine, P., 1968-80. *Dictionnaire étymologique de la langue grecque.*

Cherry, J.F., Davis, J.L. & Mantzourani, E., forthcoming. *The landscape of N. Keos in the Cyclades.*

Dahl, G. & Hjort, A., 1976. *Having herds: pastoral herd growth and household economy.*

Davies, J.K., 1971. *Athenian propertied families 600-300 B.C.*

Davies, J.K., 1981. *Wealth and the power of wealth in classical Athens.*

Delano Smith, C., 1979. *Western Mediterranean Europe.*

Ducrey, P., 1968. *Le traitement des prisonniers de guerre dans la Grèce antique.*

Dyson Hudson, R. & N., 1980. 'Nomadic pastoralism', *Annual Review of Anthropology* 9, 15-61.

Edmonds, J.M., 1957-61. *The fragments of Attic comedy.*

Ehrenberg, V., 1969. *The Greek state*, 2nd ed.

Einarson, B. & Link, G.K.K., 1976. *Theophrastus. De causis plantarum* I (Loeb Classical Library).

Finley, M.I., 1973. *The ancient economy.*

Finley, M.I. (ed.), 1973. *Problèmes de la terre en Grèce ancienne.*

Finley, M.I., 1985. *Ancient history: evidence and models.*

Forbes, H.A., 1982. *Strategies and soils: technology, production and environment in the peninsular of Methana, Greece*, Diss. Pennsylvania.

Foxhall, L., unpub. MS. 'Cato's olive grove', from her forthcoming Liverpool doctoral thesis, *Olives in the agriculture of classical antiquity.*

Foxhall, L. & Forbes, H.A., 1982. '*Sitometreia:* the role of grain as a staple food in classical antiquity', *Chiron* 12, 41-90.

Friedl, E., 1963. *Vasilika.*

Gallant, T.W., 1983. 'The Ionian islands paleo-economy research project', in Keller & Rupp 1983, 223-6.

Garnsey, P., 1988. 'Mountain economies in southern Europe: or thoughts on the early history, continuity and invididuality of Mediterranean upland pastoralism', originally in M. Mattmüller (ed.), *Itinera* 5/6 (1986), 7-29; repr. in this volume pp. 196-209.

Georgoudi, S., 1974. 'Quelques problèmes de la transhumance dans la Grèce ancienne', *Revue des Etudes Grecques* 87, 155-85.

Grmek, M.D., 1983. *Les maladies à l'aube de la civilisation occidentale.*

Halstead, P., 1981. 'Counting sheep in Neolithic and Bronze age Greece', in I. Hodder, G. Isaac and N. Hammond (edd.), *Pattern of the past. Studies in honour of David Clarke*, 307-30.

Halstead, P., 1984. *Strategies for survival: an ecological approach to social and economic change in the early farming communities of Thessaly, N. Greece*, Diss. Cambridge.

Halstead, P., 1987. 'Traditional and ancient rural economy in Mediterranean Europe: plus ça change?', *JHS* 107, 77-87.

Hammond, N.G.L. (ed.), 1981. *Atlas of the Greek and Roman world in antiquity.*
Hanson, V.D., 1983. *Warfare and agriculture in classical Greece.*
Hennig, D., 1977. 'Die Bericht des Polybios über Boiotien und die Lage von Orchomenos in der 2. hälfte des 3. Jahrhunderts v. Chr.', *Chiron* 7, 119-48.
Hodkinson, S. & H., 1981. 'Mantineia and the Mantinike: settlement and society in a Greek polis', *BSA* 76, 239-96.
Hodkinson, S., 1983. 'Social order and the conflict of values in classical Sparta', *Chiron* 13, 239-81.
Hodkinson, S., 1986. 'Land tenure and inheritance in classical Sparta', *CQ* n.s. 36, 378-406.
Hopper, R.J., 1979. *Trade and industry in classical Greece.*
Hörnschemeyer, A., 1929. *Die Pferdezucht im klassischen Altertum*, Diss. Giessen.
Hort, A., 1916. *Theophrastus. Enquiry into plants* (Loeb Classical Library).
Hughes, J.D., 1975. 'Ecology in ancient Greece', *Inquiry* 18, 115-25.
Jacoby, F., 1954. *Die Fragmente der griechischen Historiker*, Teil IIIb (Supplement).
Jameson, M.H., 1977/78. 'Agriculture and slavery in classical Athens', *CJ* 73, 122-45.
Jameson, M.H., forthcoming. 'Appendix on border disputes', in the first volume of the Argolid Exploration Project of Stanford University.
Jardé, A., 1925. *Les céréales dans l'antiquité grecque.*
Jost, M., 1975. 'Statuettes de bronze archaïques provenant de Lykosoura', *BCH* 99, 341-64.
Judeich, W., 1931. *Topographie von Athen*, 2nd ed.
Kahrstedt, U., 1954. *Das wirtschaftliche Gesicht Griechenlands in der Kaiserzeit.*
Keller, D.R. & Rupp, D.W. (edd.), 1983. *Archaeological survey in the Mediterranean area, BAR* Int. Ser. 155.
Kent, J.H., 1948. 'The temple estates of Delos, Rheneia and Mykonos', *Hesperia* 17, 243-338.
Koster, H.A., 1976. 'The thousand year road', *Expedition* 19, 19-28.
Koster, H.A., 1977. *The ecology of pastoralism in relation to changing patterns of land use in the northeast Peloponnese*, Diss. Pennsylvania.
Lamb, W., 1925/26. 'Arcadian bronze statuettes', *BSA* 27, 133-48.
Lane Fox, R., 1985. 'Aspects of inheritance in the Greek world', in P.A. Cartledge & F.D. Harvey (edd.), *Crux. Essays presented to G.E.M. de Ste. Croix, on his 75th birthday*, 208-32.
Legon, R.P., 1981. *Megara.*
Lewthwaite, J., 1981. 'Plains tails from the hills: transhumance in Mediterranean archaeology', in A. Sheridan & G. Bailey (edd.), *Economic archaeology, BAR* Int. Ser. 96, 57-66.
Lewthwaite, J., 1982. 'Acorns for the ancestors: the prehistoric exploitation of woodland in the west Mediterranean', in Bell & Limbrey (edd.) 1982, 217-30.
Lewthwaite, J., unpub. MS. 'Works and days: archaeological implications of recent settlement and subsistence activities in marginal regions of the western Mediterranean'.
Marek, C., 1984. *Die proxenie.*
Martin, R., 1973. 'Rapports entre les structures urbaines et les modes de division et d'exploitation du territoire', in Finley (ed.) 1973, 97-112.
Michell, H., 1957. *The economics of ancient Greece*, 2nd ed.
Moretti, L., 1957. *Olympionikai, i vincitori negli antichi agoni olimpici, Atti della Accad. Naz. dei Lincei, Classe di Scienze morali, stor. e filol., Memorie*, Ser. 8, Vol. 8, fasc. 2, 53-198.
Moretti, L., 1970. 'Supplemento al catalogo degli Olympionikai', *Klio* 52, 295-303.
Osborne, R., 1985a. *Demos: the discovery of classical Attika.*
Osborne, R., 1985b. 'Buildings and residence on the land in classical and hellenistic Greece: the contribution of epigraphy', *BSA* 80, 119-28.
Owens, E.J., 1983. 'The Koprologoi at Athens in the fifth and fourth centuries B.C.', *CQ* n.s. 33, 44-50.
Parker, R.A. & Dubberstein, W.H., 1956. *Babylonian chronology, 626 B.C.-A.D.75.*
Payne, S., 1973. 'Kill-off patterns in sheep and goats: the mandibles from Aşvan Kale', *Anatolian Studies* 23, 281-303.
Payne, S., 1985. 'Zoo-archaeology in Greece: a reader's guide, in N.C. Wilkie & W.D.E. Coulson (edd.), *Contributions to Aegean archaeology: studies in honor of William A. McDonald*, 211-44.
Pečirka, J., 1973. 'Homestead farms in classical and hellenistic Hellas', in Finley (ed.) 1973, 113-47.
Pippidi, D.M., 1973. 'Le problème de la main-d'oeuvre agricole dans les colonies grecques de la Mer Noire', in Finley (ed.) 1973, 63-82.

Plant, R., 1979. *Greek coin types and their identification.*
Pritchett, W.K., 1953. 'The Attic stelai. Part I', *Hesperia* 22, 225-99.
Pritchett, W.K., 1956. 'The Attic stelai. Part II', *Hesperia* 25, 178-317.
Pritchett, W.K., 1969. *Studies in ancient Greek topography*, Part II.
Pritchett, W.K., 1970. *The Choiseul marble.*
Rackham, O., 1982. 'Land-use and the native vegetation of Greece', in Bell & Limbrey (edd.) 1982, 177-98.
Rackham, O., 1983. 'Observations on the historical ecology of Boeotia', *BSA* 78, 291-351.
Renfrew, C. & Wagstaff, M. (edd.), 1982. *An island polity: the archaeology of exploitation on Melos.*
Robert, L., 1949a. 'Epitaphe d'un berger à Thasos', *Hellenica* VII, 152-60.
Robert, L., 1949b. 'Les chèvres d'Héracleia', *Hellenica* VII, 161-70.
Robert, L., 1980. *A travers l'Asie Mineure.*
Roebuck, C., 1945. 'A note on the Messenian economy and population', *CPh* 40, 149-65; repr. in *Economy and society in the early Greek world*, 1979, 1-17.
Rostovtzeff, M.I., 1941. *The social and economic history of the hellenistic world.*
Ste. Croix, G.E.M. de, 1957. Review of W.L. Westermann, *Slave systems of Greek and Roman antiquity*, in *Classical Review* n.s. 7, 54-9.
Ste. Croix, G.E.M. de, 1972. *The origins of the Peloponnesian war.*
Ste. Croix, G.E.M. de, 1981. *The class struggle in the ancient Greek world.*
Sallares, J.R., 1986. *Towards a new approach to ancient history: the interaction of biological phenomena, social structure and economy in ancient Attica*, Diss. Cambridge. (Sallares, J.R., n.d. = a subsequently revised and augmented MS version of his dissertation.)
Sartre, M., 1979. 'Aspects économiques et aspects religieux de la frontière dans les cités grecques', *Ktema* 4, 213-24.
Semple, E.C., 1922. 'The influence of geographic conditions upon ancient Mediterranean stock-raising', *Annals of the Association of American Geographers* 12, 3-38.
Semple, E.C., 1932. *The geography of the Mediterranean region: its relation to ancient history.*
Thompson, K., 1963. *Farm fragmentation in Greece.*
Van Wersch, H.J., 1972. 'The agricultural economy', in W.A. McDonald & G.R. Rapp, Jr. (edd.), *The Messenia Minnesota expedition*, 177-87.
Wagstaff, M. & Augustson, S., 1982. 'Traditional land use', in Renfrew & Wagstaff (edd.) 1982, 106-33.
Wagstaff, M. & Cherry, J.F., 1982. 'Settlement and prehistoric change', in Renfrew & Wagstaff (edd.) 1982, 136-55.
Walbank, F.W., 1936. 'Aratos' attack on Cynaetha (Polybius IX, 17)', *JHS* 56, 64-71.
Walbank, F.W., 1957-79. *A historical commentary on Polybius.*
Wallace, P.W., 1974. 'Hesiod and the Valley of the Muses', *GRBS* 15, 5-24.
White, K.D., 1970. *Roman farming.*
Whitehead, D., 1986. *The demes of Attica, 508/7 – ca.250 B.C.*
Winkelstern, K., 1933. *Die Schweinezucht im klassischen Altertum*, Diss. Giessen.
Wöhrle, G., 1985. *Theophrasts Methode in seinen botanischen Schriften.*
Wood, E.M., 1983. 'Agricultural slavery in classical Athens', *AJAH* 8, 1-47.
Woodhead, A.G., 1981. *The study of Greek inscriptions*, 2nd ed.
Xanthakis-Karamanos, G., 1981. 'Remarks on Moschion's account of progress', *CQ* n.s. 31, 410-17.
Ziessig, K., 1934. *Die Rinderzucht im alten Griechenland*, Diss. Giessen.

TRANSHUMANCE IN ANCIENT GREECE

Most students of ancient history know that in ancient Greek literature we possess two, and only two, real descriptions of transhumance proper. In *Oedipus Rex* 1121 sqq. Sophocles tells the story of *transhumance normale*: two shepherds, one from Corinth and one from Thebes, are summer pasturing their flocks on mount Cithaeron from early spring to the rise of Arcturus. Then they return to their home cities in the lowlands.[1] About five centuries later Dio Chrysostom gives a charming account (*Or.*7) of two brothers in Euboia who had been herdsman of a rich man. When he died, apparently murdered by Domitian, they withdrew to the mountains where they had their summer dwellings. This was their real homestead and when winter came and no work was available in the town, they made their huts tighter and stayed in the hills living by tilling a kitchen garden and hunting. Although the situation is somewhat irregular, it looks like *transhumance inverse*, where the more permanent houses are situated in the mountains.

If we look at the terminology which the shepherds in Sophocles use, the words *epaula* and *stathma* to the winter dwelling in the lowlands, and at Oedipus' use of the word *synaulos*, it seems obvious that *aule* was used for the summer dwelling as well. *Stathmos* is used by Dio of the summer dwellings which become permanent when the *skene* is made tighter, and the winter dwellings in the lowlands are called *skene* and *aule*. Most likely there is no great difference between the housing of men and cattle in the different areas. If we add to this the myth of Kerambos, told by Antonius Liberalis (22), we exhaust the literary evidence. Kerambos did not trust the warnings of Pan and the Nymphs that it was necessary to winter pasture in the lowlands and therefore he lost his sheep when snow came, and he was transformed into a Kerambyx. Alexander's speech to his soldiers reported by Arrian (7.9.2) has been mentioned in the same connection. Alexander reminds his soldiers that his father found them as dirty shepherds and left them as civilised Greeks who were able to defend themselves against their enemies.

This evidence has been discussed for decades by those scholars who have been banausic enough to take an interest in cattle farming. But it is astonishing in how many indices of scholarly books on Greek history the lemmata 'sheep', 'shepherd' and 'transhumance' have been omitted. In a brilliant article Stella Georgoudi collected much of the evidence.[2] She uses two sources of information; first the terminology of later Greece, where we know that transhumance was much practised; and second, the epigraphical material, drawing from Louis Robert, *Bergers grecs*,[3] but adding a lot to his rich information.

Nevertheless we have to admit that the evidence is scanty. We know that transhumance existed, but we do not know its importance for the production of meat, cheese and wool. As usual we do not possess quantitive data, to allow us to print graphs and write what is normally believed to be a real economic history. Occasionally we have a rather vague feeling of something important but nothing more.

We might leave the problem here and turn our minds to richer fields of ancient history – as most historians do. They want to stress the fact there there is no continuity in the population of Greece from antiquity to modern times and therefore, in reconstructing the past, one should avoid models derived from social anthropology. Nor do we know much about climate and the natural vegetation in antiquity. If woodland still prevailed in the mountains, there was no good summer pasture. Compared with the great periods of transhumance in Spain and Italy from the late Middle Ages where some statistical work can be done, the number of flocks in ancient Greece seems so small that it is not worth studying the problem further. Finally we are impeded by a lack of definitions: how many sheep have to be involved and how far does the shepherd have to walk with his flocks from winter pasture to summer pasture before we can dare to use the word transhumance?[4] I fear that this approach will not give us a clearer perception of the phenomenon.

The main problem is, in my view, that we do not know very much about ancient Greek agriculture at all. Two manuals – if one could call Hesiod's *Works and Days* and Xenophon's *Oeconomicus* manuals – give us very little compared with the rich information of the Latin agronomists Cato, Varro and Columella. Roman archaeology has provided us with a lot of information, whereas Greek archaeology has supplied only a few details. Only a few Greek farmsteads have been excavated and scholars still discuss whether they are farmsteads at all.[5] I shall therefore try to put the question in a different way. In Greek agriculture, where were the animals?

In the *Works and Days* of Hesiod not many animals are mentioned at all. The draught oxen are most common; comp. 405,436 sqq. and 606 sq. where the oxen and the mules are the only animals for whom winter fodder has to be collected. For an offering a heifer and milk from a goat is mentioned (590) and the gelding of sheep, pig and bull (786 sqq.). In the cold winter only sheep are protected by their wool; oxen and goats freeze (515 sqq.) – apparently they are not protected in the stable, the *sekos poimneios* (785) being merely a pen.[6] The interest in animals is almost absent but men can, by hard work, be *polymeloi* and *aphneioi*. Hesiod is rather vague on the general setting of agriculture, but he has neighbours and is certainly living in some sort of village or hamlet. No buildings are described in any detail, and it is difficult to get any precise understanding of the real world, his main interest being the *erga*. In the *Theogony* his famous calling by the muses happens when he is watching sheep on the slopes of Mount Helicon (*Theog*. 22 sqq.).

In his *Oeconomicus* Xenophon is even more silent about the animals. It seems unthinkable to breed one's own horse (3.10) and in the first discussion land and sheep are kept separate as two different parts of *oikonomia*. When Ischomachus is demonstrating the house for his young wife, nothing is said about animals or stables (8.2 sqq.) and 'the stubble may be burnt with advantage to the land or thrown on the manure heap' (18.2). No sheep are pasturing on the stubble and the manure heap is more a compost heap than anything else (20.10 sqq.). If not, Ischomachus is too much of a gentleman to use the word itself. It would in fact be a disaster to expose animal dung to rainwater, as is recommended. It is uncertain whether Ischomachus has several farms or only scattered pieces of land, but the idea that his father bought neglected farms after the Peloponnesian war and sold them when under cultivation is impossible for chronological reasons alone.[7]

This very short survey has shown that Xenophon's gentleman farmer is not interested in mixed farming. The treatise is presented in rather general terms and reveals quite a few fatal mistakes, which would have been disastrous if ever practised. Animal husbandry is passed over in silence, and the draught animals only mentioned in the description of threshing (18.3 sqq.).

If we are looking for more precise descriptions of agriculture in Greek literature, the Homeric epic is a most fruitful area. In fact nearly all aspects of agriculture are described either directly or in similes. I do not think it necessary to give a full account of the so-called Homeric Question to be allowed to use the poems as a historical source. We will never be able to establish the precise relation between the epic and the so-called historical reality. What we do know is that the Greeks themselves loved the poems and saw their own life reflected there, as in a mirror. It is a characteristic feature of the epic that everything is bigger and better than in real life. The heroes are beefeaters, not because meat was more common nourishment in the dark ages but because meat is beautiful and connected with feasting. The houses and weapons are more splendid, perhaps because they reflect the older and richer society of the Mycenean age but first and foremost because it is more heroic in that way. The similes have the function of giving life to a situation. There are lots of lions and wolves in the poems but it is not necessary to enter into a discussion of Macedonian lions or the Nemean lion. Lions were known – perhaps by literary, perhaps by iconographic tradition – to be the fiercest animal and therefore it is natural to put them into the poems.

Just as Moses Finley used the poems as a reflection of the norms of men,[8] I would use them as illustrations of their life and work, in a way as a contrast to the heroic life of the heroes. Ploughing, harvesting and winnowing were perhaps subjects for relaxation, put into a poem and presented to an alert public after a common meal, where they had their share of the meat, because the occasion was connected with an offering.

In the epic animal husbandry is often described as a struggle between herdsmen and beasts of prey. It is natural that such fighting should take place where such beasts are found, that is in the mountains, in the *kale besse* or in the *nomos*. We find the *stathmos*, *klisia* and *sekos* but no mention of farmsteads proper or cultivated fields. I find it rather surprising to read the statement of Will Richter: 'Die landwirtschaftliche Produktion der homerischen Welt hat ausgeprägten Mischcharakter, und es ist nicht leicht zu entscheiden, auf welcher Seite jeweils der Schwepunkt liegt.'[9] It is only by ascertaining that the words meaning herdsmen 'kennzeichnen nicht einen Mann, der – ausschliesslich – hütet, sondern den Mann, wenn er hütet.'[10] This may be so, but that does not permit us to talk about 'Mischcharakter' or mixed farming. The rich man has his cultivated land certainly and his flocks, but they are two different methods of investment. As far as I can see, we do not normally find domesticated animals connected with the farmstead, except, of course, the draught animals. The most normal household in the epic seems to be the palace of Odysseus in Ithaca, 'a second class palace ... where geese waddle about in the court littered with dung heaps' as Wace put it.[11] The dung heap of the palace where Odysseus finds his old dog Argos is, in fact, formed by the manure of the draught animals, mules and oxen (*Od.*17.297). In Ithaca the animal production is divided, the pig stock under Eumaeos and the goats under Melanthius in the *eschatie* (*Od.*14.104). Every day two slaves are to send their best animal to the table of the suitors. Only in *Od.*17.70 sq. the sheep are returning home, as it seems, to the palace. One wonders if this is normal practice. The simile in *Od.*10.410 sqq. does not tell about the return of the cattle to the home farm; *es kopron* could as well signify a pen where the calves are kept while the cows are pasturing further away, just like *Il.*18.475, where the cows are running to the pasture *apo koprou*.

I am, of course, here giving only an interpretation just as Richter gave his, without knowing, I fear, whether it was 'die absolute Wahrheit'. This brief examination, however, has demonstrated that agriculture and animal husbandry do not necessarily co-exist in the epic, any more than they co-exist in the writings of Hesiod and Xenophon, where there are no descriptions of cattle breeding at all. The reason for this division is not difficult to grasp. The Greeks did not cultivate fodder systematically, this being a modern invention. They could not therefore plan a system of rotation with different crops including fodder. They had to let the fields lie fallow, normally every other year. Animal husbandry was dependent on the natural grass resources, meadows near a river, fallow land during the rainy winter. And during a dry summer reserves were to be found on shrubland and in the mountains, not necessarily far away, but apart from the farmstead in the village. Odysseus had his other reserves on the mainland, if we can believe Eumaeos' (*Od.*14.100 sqq.) boasting of the twelve flocks of his master pasturing there. We do not know if Odysseus owned the land there but the swampy marshes opposite Ithaca are well adapted to large scale cattle breeding today. This is not the situation on Ithaca and in most areas of Greece.

The many exact details concerning agriculture and cattle in the epic form a strong contrast to the vague impression from later literature. An explanation might be that the focus was now on the city and the city took over offerings and monopolized offerings and feasts. Accordingly, breeding of animals was no longer closely connected with the person giving the feast and therefore no longer so interesting. And people still had the epic and listened to the marvels of old days, but herdsmen were far removed from urban life. Only in the Hellenistic period did pastoral poetry become fashionable. But it is a long way from pastoral life as depicted by Theocritus and Longus to real pastoralism. In fact, scholars studying pastoral poetry need to follow herdsmen in the mountains to understand the difference between poetry and reality. Then they would be able to explain what happens in the process of creating poetry. It is obvious, however, that Daphnis is not practising transhumance. During the wintertime his kids are fed by leaves near the village, but the pasture is on shrubland.[12]

In one situation we do hear about the animals in the period of the city state. During war it is necessary to evacuate as much as possible from the countryside. Hanson has discussed the various methods of evacuation, Athenians sending their cattle to Euboea, others sending their animals up into the mountains or collecting them in strongholds. Unfortunately the descriptions in our sources are very brief, and do not allow us to evaluate the type of cattle breeding. I shall therefore only refer to Hanson's treatment of the subject.[13] Cattle are only occasionally mentioned in lawsuits. Isaios (6.33) mentions the sale of some estates, including a shepherd and a flock of goats, for 13 minae. It is remarkable that this item is sold separately and there is no connection between the flock and the land. In 11.41 he mentions another property consisting of land near Eleusis, 60 sheep, 100 goats, furniture, a fine horse etc., again mentioned separately. In Demosthenes (47.52 sqq.) a man claims that his opponent stole 50 sheep and the shepherd with his utensils. Here too the situation is not quite clear, but the shepherd, presumably also a slave, and the flock are surely a unity kept separate from the land.

I shall not go into further details. But in my view it is obvious that, as far as we can see from the literacy sources, animal husbandry and agriculture were considered by the Greeks to be separate. This may be due to the settlement patterns. I shall not here enter into the discussion of single farms or village habitation. Both types are mentioned in the sources and many variations existed. It was, however, difficult to keep a large number of cattle within the city walls and, when necessary during wartime, it occasioned trouble.[14] The latest contributions to the discussion of rural settlement tend to minimize the numbers of single farms.[15] Two small excavated villages have been interpreted as summer dwellings for transhumant herdsmen. Karphi above the Lasithi plain in Crete situated ca.1100 metres above sea level and inhabited c.1200-800 B.C. is discussed by L. Wance Watrous,[16] and a site near the village Vitsa in the central part of the Pindos range, inhibted c. 800-400 B.C. is published by Sotirios Dakaris.[17] The structure of the villages suggests transhumance inverse, where the winter pasture was centred on less permanent dwellings in the lowland. A systematical archaeological survey of remote parts of Greece might well give more archaeological evidence.

Finally, if one focuses on the borderland between city states a common pasture is often found. We have already noticed Mount Cithaeron mentioned by Sophocles and we could add Thuc 5.42, where we are told that *palaioi horkoi* existed between the Athenians and the Boiotians concerning Panactos. This area had to be left as a common pasture for the two people and they were not allowed to build there. Of course, the details are only mentioned because they gave rise to troubles in wartime. A similar no man's land existed on the Mount Parnassus and caused the outbreak of the Corinthian War (*Hell.Ox*.8.3). Apparently no treaty existed between the Locrians and the Phocaeans. This text is discussed by Georgoudi together with several laws and treaties concerning *epinomia*, the right for members of different communities to grass on common pasture. Another problem is the right to pass with the flocks from one territory to another. This was also regulated by laws and treaties and discussed in the same connection. I shall not repeat her arguments but only refer to her scholarly treatment.

Summing up, I must stress the non-sedentary character of animal husbandry, the existence of remote common pastures and the existence of rules regulating the traffic of the flocks. All this bears witness to transhumance. As geographical conditions vary very much over short distances in Greece proper, the wanderings of the flocks were not necessarily long. The landscape is rather different from that in Italy where the sheep pasture in winter in the southern plains of Apulia and in summer in the Apennines.

Whenever cattle are mentioned in connection with the sale of property and the like they are mentioned especially, not as being a part of landed property. Greek agriculture, it seems, is not described in the sources as mixed farming in any technical sense. It has been the view of most scholars – from Jardé to Michell and P.J. Hopper[18] – that this is the reflection of the rural reality, the only lonely voice being Will Richter, who, in *Archaeologica Homerica* speaks of the 'Mischcharakter' of the 'Wirtschaftliche Produktion'. If by this he has in mind some type of mixed farming, I disagree. Mixed farming should in my opinion be used when we are speaking of the ecotype where both livestock and crops are raised for the commercial market, perhaps better called 'balanced livestock and crop raising' to follow the definition of Eric Wolfe, opposed to the so-called Mediterranean ecotype.[19] However, since the problem of mixed farming has been raised again in the paper of Dr Hodkinson, I shall make some comments on this topic. The problem is that there are serious differences in our interpretation of the nature of Greek agriculture.

Hodkinson uses two models, which I shall cite in his words.

> The first, and older model which draws its inspiration from observations of traditional patterns of agriculture in recent centuries, assumes a large measure of divorce between arable and pastoral elements [and]
>
> an alternative model .... which questions the applicability of recent traditional agricultural practises to the ancient past, and argues for a fair degree of symbiosis between arable culture and animal husbandry. It emphasizes the rotation of cereals and legumes in arable cultivation, with a consequent increased availability for summer fodder in the lowlands, thus evening out seasonal imbalances and diminishing the need for transhumance.

The title of his paper is 'Animal Husbandry in the Greek Polis', and it is therefore natural that he concentrates on the second model. As my subject is transhumance proper, it is obvious that I should concentrate on the evidence for the non-sedentary character of Greek animal production, the existence of remote animal pastures in the mountains and rules regulating the traffic of flocks. But I do not exclude the possibility that a small amount of cattle, especially sheep and goats besides the draught animals, was kept by the ordinary Greek peasant; this means the man of hoplite census tilling perhaps about five hectares of land.[20]

In his examination of the evidence of transhumance Hodkinson underlines that this type of cattle rearing was especially practised by the rich people. Here I agree. Investing in flocks is dangerous as is the life in the mountains. But I would add that arable land in the city state was limited and therefore one should expect that the natural reserves in the mountains were used by the group who could afford it and who dared run the risk.

I find it more serious that Hodkinson maintains that the Greeks did grow quite a few fodder plants and practised a regular rotation of crops. We know that the alfalfa, *poe medice*, was introduced in the fifth century, but it is not frequently mentioned and we do not know if it was in common use. I doubt it. Luzerne has to be left for several years to give its best yield, and I do not think that the average farmer had the acreage to leave this plant for six or seven years, if he was trying to be selfsufficient with the most necessary products, corn, olive and wine, and eventually have a little surplus to sell.[21] I think the alfalfa was normally reserved for the horses, where this animal of luxury was bred. Neither is it probable that cytisus, *medicago arborea*, was very important, although the plant was sown as well.

As to the other forage plants mentioned by Theophrastus I still doubt their importance, differing here from Hodkinson's evaluation of the great botanist as a source for Greek agriculture. I do not believe that he was interested in agriculture as such, but as a botanist he pays due respect to cultivated plants. He does not mention farm work much, and, while he is in some respects our best source, it is because of the general lack of evidence. He makes excellent remarks on the quality of soil and the different types of grafting and propagation but he never gives advice as to which of the methods should be preferred. The work of vintage and harvest are never described, obviously because they are beyond the systematical framework of his enquiries into plants.

When Hodkinson argues that the many varieties of fodder plants mentioned by Theophrastus bear witness to their common use, one could maintain that any agrarian society in fairly dry climates must rely on every fodder supply available, including the dry remains of threshed pulses. When, for instance, the peasants from the Island of Lesbos today from the month of September carry olive branches with leaves home to the village it is obviously because this is one of the last resorts before the winter rain. Although Theophrastus mentions a lot of plants which could be used as fodder, he very seldom discusses their role in the agrarian economy. One of the very few places where he refers to alfalfa, he adds that some say that the crop is destroyed by the urine and dung of sheep. This does not imply great familiarity with the cultivation of this plant. Theophrastus is not characterized as a good agricultural writer by Varro, who says that his books are not as useful for those who wish to tend the land as for those who wish to attend the schools of philosophers (*R.R.*1.5.2) – a view with which I do not agree.

The problem is how many animals the farmer could keep near his homestead. Here I do not think that any norm can be established, as the conditions vary greatly from area to area. Our literary sources, of course, often refer to the situation in Attica, one of the most arid areas in Greece. Lin Foxhall in a recent article says that in Methana – even more dry than Attica with an average of about 400 mm of rain per year – most households own a 'small flock of lean, agile, coarse fleeced sheep and goats, usually not more than six animals.'[22] Even in Mantinea in Arcadia Hodkinson mentions 'the ownership of a small number of sheep, pigs and goats' only, although the average rainfall is just over 800 mm per year.[23] On the island of Ceos only a few sheep – *oliga probata* – were stall fed with cytisus, leaves of figs and olives – because the inhabitants had no pasture land, *nomos*. This is said by Aelian (*HA* 16.32) to be something extraordinary, the norm being that the sheep were fed on the *nomos*.

Stall feeding was not the normal practice. The peasants used shrubland and pastureland often separated from the arable land, and of course stubble and fallow land if the stubble was not burned, as recommended by Xenophon. Our sources testify to a fairly small scale of animal production linked to agriculture and I would interpret not a few examples of flocks mentioned by the Attic orators as non-sedentary animal husbandry. Flocks are sold, even with the shepherd, apart from the land. We should, therefore, not be too sure that we have an example of a 'fair degree of symbiosis between arable culture and animal husbandry.'[24]

Finally one should ask the question: What does it matter if small numbers of animals were living on the farmstead or near the farmstead but apart from it? It is of vital importance because of the manure. If the animals were stall fed, as in Keos, one could use the manure entirely; if the sheep and goats graze on stubble and fallow land, the sheep are the best manuring machine, as K.D. White calls them.[25] But if the flocks live and graze most of the day on pastures or shrubland, they will drop the manure on land which is not under culture, a waste from the farmers point of view.

Hodkinson has tried to press the sources for all possible information concerning forage plants, and, if we accept his interpretation, it would give a larger amount of manure and a possibility of intensifying agriculture interalia by not using fallow every other year, as most of our scanty sources testify. Of course, the Greeks grew legumes on fallow land when possible, trying to shorten the fallow period. But can we say that they had a rotation system: corn, legumes, fallow? I doubt it. They did not recognize the nitrogen deposited by their roots, since Xenophon and Theophrastus recommend the ploughing in of the whole plant of green manure. But then, of course, the value as forage plants disappears. Incidentally I have never understood how one can plough in green manure with an ard! Theophrastus discusses the burden of the different legumes on the soil and does not seem aware of the fact, as are modern peasants, that the presence of these plants alone was manure for the soil.[26]

Much as I admire a bold attempt to reanalyse Greek agriculture, I must confess that I am not convinced by the second model, a sort of mixed agriculture in a more modern sense. I disagree in the evaluation of Theophrastus and I disagree in the interpretation of not a few of the texts referred to by Hodkinson.[27]

If a *via media* needs to be found between the two models,[28] I would suggest that there were only ever a limited number of animals. If the demand was slack because the Greeks only used animals for offerings, one might ask if there was as much offering as often assumed. We should bear in mind that the sheep especially were necessary because of the wool. According to Halstead one should calculate about four to ten sheep to answer the demand of clothing pro persona based on a modern estimate,[29] which might be refined if one studied closer the dressing of the ancient Greeks.

Wool is particularly interesting, as sheep are sheared before going uphill; that is, when winter pasturing is the lowland near the cities and the market. Therefore the problem of transport is minimized. The famous Milesian wool was certainly collected in this way. One should bear in mind that Greeks might very well have been involved in transhumance all over the Mediterranean world, among other places in southern Italy and southern France, where the wool trade is better attested. In this paper, however, I have concentrated on some of the problems on the Greek mainland knowing that our literary sources are concentrated in areas where transhumance presumbly played a minor role. Other areas distant from the central cities, e.g. in Arcadia and Messenia, probably had a much more pastoral character, and therefore they have left no trace in the literary sources.

Acknowledgements. I should like to thank Dr Hodkinson who has kindly given me the opportunity of seeing the final draft of his article in this volume. Although in some cases he has incorporated details from our discussion in Bern in his treatment, we are still offering two different interpretations of Greek agriculture. I have chosen to be brief, which is easier for me, since I am to some extent defending 'the older and more conservative view'. But I do not think that it is necessary to go into further details, especially as I hope to evolve my views on Greek agriculture in a forthcoming book in collaboration with Mrs Signe Isager, of the University of Odense. I should like to thank her for stimulating discussions and excursions in the Greek countryside, and also Hamish A. Forbes and Lin Foxhall for hospitality and help in Methana.

UNIVERSITY OF COPENHAGEN JENS ERIK SKYDSGAARD

NOTES

1. M.M. Austin and P. Vidal-Naquet, *Economic and social history of ancient Greece* (1974) 286-87.

2. 'Quelques problèmes de la transhumance dans la Grèce ancienne', *REG* 87 (1977) 155-85.

3. *Hellenica* 7 (1949) 152-60, *Hellenica* 10 (1965) 28-33.

4. P.D.A. Garnsey, 'Mountain economies in Southern Europe', *Itinera* (1986) 7-29 (reprinted here pp. 196-209), gives such objections!

5. J.E. Skydsgaard, *L'Agricoltura Greca e Romana:* Traditioni a confronto, *ARID* 16 (1987) (forthcoming).

6. '*boas endon eontas*' (452) does not necessarily mean that the oxen are stabled.

7. Discussed by V.D. Hanson, *Warfare and agriculture in classical Greece* (1983) 141.

8. M.I. Finley, *The World of Odysseus* (1954).

9. *Archaeologia Homerica*, ed. F. Matz und H-G. Buchholz, II, H, 'Die Landwirtschaft im homerischen Zeitalter' (1968) 32.

10. ibid. 35.

11. Wace and Stubbings, *A companion to Homer* (1963) 489, referring, it seems, to the geese of Penelope, *Od.* 19.536 sqq.

12. Longus, *Daphnis and Chloe* II. 20.

13. Hanson (n.7) 87-98.

14. Hanson (n.7) 97.

15. G.D. Rocchi, 'Gli insidiamenti in villagi nella Grecia del V e del IV sec.a.C.' *MIL* 36 (1981) 325-86. For Attica see especially R. Osborne, *Demos, the discovery of classical Attika* (1985). The epigraphic evidence is discussed by R. Osborne, *BSA* 80 (1985) 119-28.

16. *Aegean settlements and transhumance*, Temple University Aegean Symposium, 2:2-6 (1977).

17. 'Von einer kleinen ländlischen Ansiedlung des 8.-4. Jahrh. zu einer spätklassischen Stadt in Nordwest-Griechenland', *Palast und Hütte*, ed. D. Papenfuss und V. Strocka (1982).

18. A. Jardé, *Les céréales dans l'antiquité Grècque*, (1925) *passim*; H. Michell, *The economics of ancient Greece*, ed. 2. (1957) 59; R.J. Hopper, *Trade and industry in classical Greece* (1979) 147-63.

19. E. Wolfe, *Peasants* (1966).

20. A. Burford Cooper, 'The family farm in Greece', *Clas. Journ.* 73 (1977/8) 162-75, with references.

21. The situation in antiquity seems very similar to the situation in modern Greece described in *Pasture and fodder development in the Mediterranean countries*, Technical assistance mission no. 56, publ. by OECD, (1951) 53-67.

22. *History Today*, 36 (July 1986) 36.

23. S. and H. Hodkinson, 'Mantineia and the Mantinike, Settlement and Society in a Greek Polis', *BSA* 76 (1981) 239-96.

24. See Hodkinson in this publication; cf. Burford Cooper (n.20) 172: '50 prize sheep were carried off with their shepherd while they were grazing. Where had they been? The speaker says, "the thieves, not content with having the sheep, went on to my farm..." (Dem. XLVII, 52-57). The sheep were obviously somewhere else, on public pasture'.

25. K.D. White, *Roman farming* (1970) 134.

26. *H.P.* 8. 9.

27. A few examples will suffice. Hodkinson has drawn on a lot of references to cattle, esp. sheep, for instance in the writings of Xenophon. I cannot use this material because the nature of cattle rearing cannot be established. Neither can I see that Thucydides 2.14 and 7.27.5 confirm 'that the agro-pastoral farm was not merely an upper-class phenomenon in Attica'. The two texts say that the Athenians evacuated their livestock but *oi Athenaioi* does not mean 'the mass of Athenian farmers' but the Athenians as opposed to the invaders. Neither can I interpret Xenophon's descriptions of the estate in Scillous as Hodkinson. No arable land is mentioned by Xenophon (*Anabasis* 5.3.9-11), only some cereal-products, and the lovely descriptions seem to me to be one of the niches in the Greek landscape consisting of a river valley with meadow surrounded by hills with wild trees and game and

an orchard near the temple. When Pausanias visited the place, it had retained its character (5.6.4: 'Scillous is stocked with wild pig and deer for hunting, and the river runs through this countryside'. Finally I think that the argument from Xenophon, *Oec.* (5.3) that the *techne* of animal husbandry (*probateutike*) is linked with that of tillage (*georgia*) is weakened when one realizes that the link is 'that men have victims for propitiating the gods with sacrifices and cattle for their own use'. The discussion of the relation between *agricultura* and *pastio* is repeated by the Roman agronomes, e.g. Varro *R.R.* 1.2.13 sqq. with many references to the ancient literature now lost. Agriculture and cattle rearing seems normally to be treated as two different subjects in the technical literature of antiquity. Does that not reflect the rural reality?

28. The word 'model' seems a little misleading when applied to the results of the long tradition of scholarship leading to the 'old view on Greek agriculture'. The interpretation given by Semple, Jardé and others has long passed the stage of 'being a heuristic device for structuring a discussion'. The 'new model' stems from interpretations of archaeological remains in prehistorical Thessaly and cannot, in my view, be applied directly to classical Greece.

29. P.L.J. Halstead, 'Counting sheep in neolithic and bronze age Greece', *Patterns of the past; studies in honour of David Clarke*, ed. Hodder, Isaac and Hammond (1981), 327-28.

SACRIFICE AND ANIMAL HUSBANDRY IN CLASSICAL GREECE

Introduction

The magazine *The Economist*, for September 3rd, 1983, reported that at Mecca, in two weeks' time, Abraham's substitution of a ram for the sacrifice of his son Isaac was to be 'commemorated by the slaughter of 1,200,000 animals flown in from Australia, Bulgaria, Turkey, Argentina and Uruguay and killed on behalf of 2 million Haj pilgrims. This creates a sudden surfeit of meat. Most is burned or buried but efforts are being made to find a way of using it.' I cite this article on some of the practical consequences of animal sacrifice in the present day as a reminder of the powerful impact religious belief and practice can have on the use of animals in traditional societies (cf. Dahl and Hjort, 1976: 162-3). Sheep and goats were the most numerous victims in classical Greece, as they were the dominant livestock in the countryside. But we need to go beyond the limits of the pastoral theme of the conference which has called forth the papers in this volume and include all that is covered by animal husbandry. To understand the information from the sacrifice of any one type of victim we need to consider the alternatives open to the Greeks, which were (aside from birds) cattle, sheep, goats and pigs. So too for the economy of the Greeks pastoralism has to be seen in relationship to alternative uses of the land, that is to animal husbandry as a whole and to agriculture. (See Hodkinson's paper in this volume.)

The Greeks derived virtually all their meat (aside from game) from the ritual of sacrifice.[1] Imperfect animals were normally regarded as inappropriate but the standards by which they were rejected would have been a matter of judgment (Stengel, 1910: 197-202; Ziehen, 1939: 589). The Spartans, known for their frugality, were said to sacrifice lame victims ([Plato], *Alc.* II 149a) and those who had little alternative no doubt killed and ate animals unsuitable for ritual. Working cattle were prohibited for sacrifice (e.g., Aelian *V.H.* 5.14) and the rare, explicit requirement of them in sacrifice was for the purpose of making a distinct symbolic statement.[2] But old plough- and draught-oxen may also have been consumed privately by the poor. Xenophon, when he had no alternative, sacrificed a cart ox for divination (*Anab.* 6.4.25). An absolute ban on the sacrifice or killing and eating of weak, sick and old animals would have had a marked effect on the returns obtained from herding but I do not think we are required to suppose this was the case.[3] But even if marginal members of the community sacrificed less desirable specimens and killed and consumed old working animals without ceremony, the great bulk of the meat consumed by the classical Greeks would still have come from sacrifice, directly or indirectly (from butcher shops that purchased meat from sacrifices).

In sacrifice of the normal type (which the world *thysia* usually refers to, sometimes misleadingly termed 'Olympian sacrifice' by modern scholars) very few usable parts of the animal were destroyed in the fire (on the division of the victim between gods and men, cf. Hes. *Theog*. 535-612 and Puttkammer, 1912). In ritual calendars few rites demanded holocaust, the complete destruction of the victim. For example, in the Attic deme of Erchia, in the first half of the fourth century B.C., out of 56 victims (all sheep, goats or pigs) only one ewe lamb and three piglets, costing altogether 16 drachmai out of a total of 520 drachmai, were holocaust; in addition, the skins of two does (female goats) were to be torn apart, if the language is understood correctly. All else was at the disposal of the worshippers (*LSCG* 18).[4] The Greeks did in fact employ a wide range of powerful actions, including rites referred to as *sphagia* and *enagismata*, that involved the destruction for particular purposes (e.g., oaths, purification, aversion, communication with figures of the underworld, fertility magic) both of customary victims and of other animals, such as dogs and horses, not eaten in the classical period.[5] Except for the very small pigs frequently used in such rites (discussed below), the animals killed in these powerful actions were much fewer in number than the victims of normal sacrifice and were of little importance for animal husbandry and the food supply. Thus a description of Greek sacrificial practice is in effect a description of Greek procurement of meat for consumption and of part of their supply of animal skins, indispensable for many purposes. (See Appendix.)

The main questions we can ask of the information obtained from sacrifice are, first, what regime of animal husbandry is indicated by the pattern of sacrificial slaughtering (i.e., the species, sex and age of animals killed and the relative proportions)? Secondly, what was the contribution of sacrifice to the diet? Finally, did the needs of religion interfere significantly with the other goals of animal husbandry (such as the production of milk, wool, hair, leather and manure, and the use of cattle for work)? The evidence from inscriptions and from archaeology to be used here is presented first, followed by a discussion of cattle, pigs, sheep and goats. We conclude with a consideration of diet and of the impact of religion.

Before examining the details it may be useful to summarise the three principal strategies that have been distinguished for the maintaining of herds:

(1) The aim is to have animals for work and for the production of wool, hair and hides. There is equality among male and female animals with most of the males being castrated. (Male sheep produce more wool and wethers are easier to control than rams; on the preference for the flesh of castrated animals, Galen *De semine* 1.15, IV, p. 573 Kühn.) There is limited breeding for the purpose of

maintaining the herds rather than increasing their size and selling off what cannot be run on the land available. Animals tend to be kept to maturity.

(2) The aim is milk production. The herds are mostly female. There is a high reproductive rate and most young males are killed. Females tend not to be kept beyond twelve years for cows, six for ewes and does, and sterile females of all ages are culled.

(3) The aim is meat production. Young males are killed, sometimes after fattening, as juveniles or young adults, at the point when the most economical balance has been achieved between the weight of the animal and the feed consumed. The mean age of animals slaughtered is less than for (1) or (2). Pigs, unlike the ruminants, are raised only for this purpose.

In most situations more than one of these aims are desired and actual practice is a compromise between various strategies.[6]

The evidence from inscriptions

For the evidence from sacrifice I draw primarily on inscriptions from three Attic demes. We do not have the full ritual calendar of any one community but these three, of the fifth and fourth cent. B.C., offer a good approximation. They allow us to see the relative importance of different victims, and the times of year at which they are offered make them more useful for our purpose than information on any one festival. The sacrifices prescribed are only those for which the deme

Table A Minimum number of animals sacrificed in any one year

	cattle	rams	wethers	ewes	lambs	bucks	does	kids	pigs	piglets	total
MA	6	1	9–12	9–12	—	1	—	—	2 preg. sows	7–9	35–43
			all sheep: 19–25						alt. years		
TH	2	—	16	9	3m/1f	2	2	2m	—	2m/3f	42
			all sheep: 29			all goats: 6					
ER	—	1	19	10	1m/3f	5	6	1m	—	10	56
			all sheep: 34			all goats: 12					

Notes: No restorations are included. Cattle and piglets are not usually identified by sex unless, for cattle, a pregnant victim or a bull is required; prices do not provide a clue. There was a general tendency to offer male victims to males and females to females but it was not a firm rule (cf. Paus. 9.19.7). Besides the solitary goat in the Marathon calendar there are five detectable in the fragmentary list of sacrifices made by the whole Tetrapolis (col. A of *LSCG* 20). The Thorikos calendar includes a *trittoia* but the three animals composing that sacrifice varied. (MA = Marathon, TH = Thorikos, ER = Erchia.)

officials are responsible; those rituals in the same area for which the state, cult groups or kinship groups were responsible are not listed, and of course private sacrifices made for personal reasons are beyond our ken. One of the calendars is for the deme of Marathon (*LSCG* 20 = *IG* II^2, 1358). It lacks part of the first quarter of the Attic year; it also shared in the ritual calendar of the organization of four communities known as the Marathonian Tetrapolis, which evidently saw to setting up the inscription that carried the joint and separate calendars. For the relatively large but parsimonious deme of Erchia we have a virtually complete calendar that represents the greater part of the deme's ritual obligations (*LSCG* 18 and note 4, above). The most recently discovered stone is thought to come from the deme of Thorikos. The first month of the year is mutilated.[7]

The archaeological evidence

Archaeological evidence from the remains of animal bones found in sanctuaries promises to become a valuable source of information as the techniques of excavation and analysis used on prehistoric sites are applied to sites of historical periods (cf. Payne, 1985). Hitherto, however, although there have been frequent references in publications to layers of ash and to bones, few analyses have been made and only preliminary reports have appeared for most of those that are now being undertaken. Furthermore, as with a *lex sacra* for an individual cult, even the most informative report from a single sanctuary cannot by itself show the relative importance of the slaughter to which the bones testify. I summarise what is most useful for our purposes, omitting details primarily of interest for the study of ritual.

Kabeirion, near Thebes in Boiotia. Classical to Roman (the latter mostly ca A.D. 50 to 150). Minimum number of individuals: cattle 25, sheep/goat 125, pig 7. By period, the number of identifiable bones, excluding astragali (knuckle-bones, see note 8), were:

Classical: cattle 43, sheep/goat 247, pig 17.
Hellenistic: cattle, 121, sheep/goat 504, pig 12.
Hellenistic-Roman: cattle 25, sheep/goat 153, pig 1.
Roman: cattle 285, sheep/goat 1,136, pig 15.
After A.D. 350: cattle 4, sheep/goat 17.
Overall, cattle 15-20%, sheep/goats 75-80%.

Among sheep and goats there was a preference for juveniles and subadult–adults over either infants or adults, except in the Hellenistic period when the

Table B Prices of victims in drachmai

	Marathon	Erchia	Athens (State Calendar)[1]
ox or cow *bous*	90[2]	—	—
calf, *bous leipognōmōn*[3]	—	—	25
pregnant pig *hus kuousa*	20	—	—
ram *krios*	12	10 (?)	17
pregnant ewe *ois kuousa*	16	10	—
buck *tragos*	15	—	—
wether *ois*	12	12	15
ewe *ois*	11	10	12
goat, male *aix*	—	12	—
doe *aix*	—	10	—
lambs *arnes, amnai*	—	7	4 *ois leipognōmōn*
kid *eriphos propt.*[4]	—	5	—
suckling pig *choiros*	3	3	3

Notes: (1) From *LSCG, Suppl.* 10, the largest of a number of fragments of the revision in the late fifth and early fourth cent. B.C. of the Athenian state calendar of sacrifices; some other fragments are *LSCG, Suppl.* 9 and *LSCG* 17. Thorikos did not regularly list prices, aside from two oxen between 40 and 50 drachmai. (2) One entry for Marathon, *bous* apparently followed by 150, is of uncertain significance. (3) On the meaning of this term, Hansen, 1973, who argues that it refers to an animal in the first few weeks of life. (4) On the abbreviation, presumably for *proptorthios*, referring to the age of the kid, see Daux, 1962: 626–7.

---

distribution is somewhat more even. Infant and juvenile animals as opposed to subadults and adults, except for the poorly represented classical period, constitute 1/3-2/3 of the total.

Sex is not listed. Aside from a possible disinclination to offer pigs there seems no preference for a species in this cult. The actual weight of cattle bones exceeds that of sheep/goat bones, 14.475 kg to 13.800 (among which identifiable goat bones weigh 2.27 kg, sheep 1.69 kg), an indication of their relative contribution despite the lesser number of individuals. Boessneck, 1973.

Sanctuary of Demeter at Knossos. Late fifth cent. B.C. to second cent. A.D. (in an earlier phase, ca 700 B.C., the site was not certainly a sanctuary). In the Classical, Hellenistic and Roman phases the percentages of pig bones (not minimum number of individuals) were 91, 94 and 91.8, respectively, while sheep/goat were 6.0, 4.7 and 5.2. Other bones were of insignificant number. Pig was a common victim for Demeter (Stroud, 1965: 22; Kadletz, 1976: 126-31). By contrast the Late Geometric phase had only 16.7 pig to 54 sheep/goat and 16.7 cattle, and Minoan levels under the sanctuary and Minoan-Roman deposits elsewhere in Knossos showed a similar distribution: pig 29.8 and 24, sheep/goat 62.1 and 57, cattle 6.3 and 15. Information on age, sex and parts found has not been published (Jarman, 1973).

Artemision at Ephesos. Seventh to fourth cent. B.C. Minimum number of individuals represented: cattle 8, sheep 6, goat 39, pig 6. The cattle are described as subadult. Few goats were younger than nine months and most were at least a year old, as was also the case with sheep. The pigs were juvenile. As for sex, where identification was possible, cattle were both male and female, sheep female and of the goats there was one buck and seven does (Wolff, 1978). A large collection of goat horn cores (both left and right) are explained as coming from an altar composed of horns, as known at Delos (from literature and inscriptions) and from excavation at Dreros (Marinatos, 1936), Kato Simi on Crete (Lebessi, personal communication) and Halieis (see below). Goat was a common victim for Artemis (Kadletz, 1976: 87-96).

Kommos, Crete. 1000 B.C.-A.D. 150. 21,000 bones, burnt and unburnt, the latter mostly from altars and hearths and consisting mostly of hindquarters of sheep or goats. One altar (U of Temple B2-3) had cattle only, and all parts of the animal. Pig bones were found mostly in Temple C and were rarely burnt. (Reese, 1984). The full publication of this material by Joseph Shaw, the excavator, and David Reese, who is studying the bones, will be exceptionally valuable because more than one ritual and a long stretch of time are represented at the site.

Altar of Aphrodite Ourania, Agora of Athens. Fifth-fourth cent. B.C. Over 95% of the faunal remains were burnt. The great majority of the bones are sheep/goat (at least 79 individuals), not older than three years and 77.2% under 3-6 months old.

No burnt cattle but some pig and fish bone have been identified (Reese, forthcoming).[8]

Temple of Apollo at Halieis, Argolid. Fifth cent. B.C. (but with possible earlier accumulation). In the middle room of the Early Archaic temple, entered from a door in the side of the building, 41 goat horn cores, from at least 23 individuals aged two years or older, may have constituted an interior altar of horns, as at Dreros on Crete (Marinatos, 1936). From the middle and back rooms of the temple were infantile (one to two months old) sheep/goat tibiae, metatarsals and pelves from a minimum of 12 individuals. The bones are primarily from the right side (10:1, 9:2, 12:2). The tibiae of very young pigs are also represented (at least 49). The bones do not seem to have been burnt but light burning cannot be detected because of the effect of long immersion in the sea (the temple became inundated early in the first millennium of this era). The bones, other than the horn cores, appear to have been the remains of priests' perquisites, whereas the horn cores came from sacrifices of larger animals (perhaps carried out on the altar in front of the temple which may also have had a core of horns). The absence of bones from larger animals in the middle and back rooms may be an indication of the relative rarity of adult animals in sacrifice since there is no reason to suppose the priests would forego their usual perquisites. (Alternatively, the more substantial *gera* may have been taken home for family use or for sale.) (Information on the bones from John Watson, personal communication. For the temple, see Jameson, 1973-74, 1974a and b.)[9]

Cattle

Cattle were the major sacrificial animals of antiquity and were the predominant subjects of the typical sacrifices depicted in myth and in art. Fifth-century Athens required symbolic acknowledgement of her position from her colonies and members of the Delian league by the contribution of a sacrificial cow to the Panathenaia (*IG* I³ 34, 42 = ML 46; I³ 46, 11 = ML 49; I³ 71, 57 = ML 69). For the heroic world they were a unit of value and a store of wealth and the Homeric heroes are constantly shown sacrificing and eating cattle. The proportion of cattle to other animals in Dark Age settlements is likely to have been higher than in either of the relatively more populous periods that preceded and followed. ('Animals, domestic or not, are most plentiful precisely in countries still relatively untenanted by man.' Braudel, 1973: 133.) At Dark Age Nichoria it has been estimated that they constituted 63% by weight of the protein in the diet, compared to 40% average for all other periods (Middle Helladic, Late Helladic and Byzantine). The mean age at slaughter drops from 10 to 5.5 years, an indication of an economy more concerned with meat than milk production

(Sloan and Duncan, 1983: 64-66). The number of cattle is a function of the amount of good land for pasture available near the settlement site at a time of low population pressure.

The relative scarcity of cattle in the Greek landscape today may lead one to underestimate the importance of cattle in the past, especially in comparatively underpopulated periods. In the southern Argolid ca A.D. 1700 there were some 700 cattle for a population of about 2,000. It is evident that this notably arid environment was capable of keeping a large number of cattle per person, in addition to 17 times as many sheep and goats. Two regimes can be detected – the maintenance of plough- and draught-animals in the lowland villages near the coast, and the breeding of animals, probably for both work and meat, in inland and upland areas. By 1911 cattle had declined to 512, whereas sheep and goats had increased by almost half.[10]

But for all the traditional importance of cattle as symbol and their desirability as a source of meat, their relative numbers drop as the human population increases, although their value for work in the fields, for manure and for hauling rises. After the Dark Age most sacrifices in most parts of Greece were not of cattle. The term *hiereion*, 'sacrificial victim', when used without further specification, refers to sheep (e.g., *LSCG* 88 = *SIG*³ 1039) an indication that they were the most common victim.

The local Attic *leges sacrae* suggest what was acceptable for cult and normal for consumption when the wealth of the Athenian state was not available to pay for victims. The demes, organizations with modest resources, offered few cattle compared to sheep, goats and piglets (Table A). At Erchia no cattle at all were sacrificed. At Marathon at least six cattle were killed each year, compared to 29-37 other victims (a ratio of 1 : 5-6), but the cattle cost no less than 600 drachmai compared to less than 400 for the other victims (the range comes from the fact that there was variation in the victims offered in alternate years). Thorikos offered two oxen compared to 35 sheep and goats and five pigs. The *genos* of the Salaminioi, which could afford five 40-drachmai pigs, offered only one ox in the course of its year (*LSCG*, *Suppl.* 19 = *Hesperia* 7 (1938) 1-68). In no case is it evident that cattle were ritually required although the sex and the pregnancy of a victim could be significant. For economy's sake a large sacrifice might be offered only once every other year. Thus at Marathon Athena Hellotis received a cow, three ewes and a piglet one year, but only a single ewe the next year, in which (Demeter) Eleusinia received a cow. Cattle were larger sacrifices, more pleasing to the gods and more rewarding to men. The choice between cattle and sheep (goats have a more restricted function) was essentially a matter of cost and of availability.

The evidence of the bones does not contradict that of the Attic calendars, whose information has been summarised above. At the Theban Kabeirion the ratio of cattle to all other domesticated animals is 1:4-5 for all periods, 1:6 for the classical (Boessneck, 1973; bronze and lead figurines of cattle are conspicuous at the sanctuary, Roesch, 1985). At the Ephesian Artemision cattle are 1:6.5 when compared to all other animals (minimum number of individuals, Wolff, 1978). At Halieis and the Athenian Altar of Aphrodite Ourania traces of cattle are respectively, negligible and absent. Probably at the Artemision and certainly at the sanctuary of Demeter at Knossos (Jarman, 1973) the figures are skewed by the cultic preference for certain victims. But the overall effect of many cults with other preferred victims is to free the worshippers from having, regularly, to make this large and costly offering.[11]

If, however, the contribution of cattle in weight (and therefore protein) is considered, the disproportion between them and all other victims is considerably reduced. The following have been proposed for the live weight and the usable meat (approximately one half) of adult animals at Dark Age Nichoria (Sloan and Duncan, 1983: 76): cattle 200/100 kg., sheep 40/20 kg., goat 60/30 kg., pig 100/50 kg. The relative weights rather than the precise figures are significant here. If for Marathon we suppose all animals are full-grown except for the piglets, for whom a weight of 3 kg. is used (there seems to have been a rough equivalence of 1 drachma per kg. of meat of the live animal) the meat contributed by cattle comes to 600 kg., that of the maximum number of all other animals to 615 kg. (6 cattle at 100 kg. and 25 sheep at 20 kg., 1 goat at 30 kg., 1 pig at 50 kg. and 5 piglets at 3 kg.). The fact remains that Marathon, with no lambs or kids, in contrast to Erchia with no cattle, is tilted towards the pattern of the more extravagant state cults while Erchia may represent better the situation in most smaller communities. Local conditions are relevant. There was good pasturage for cattle in the plain (Paus. 1.32.7),[12] and it was the Marathonian bull that Theseus captured and brought to the Acropolis to sacrifice, a story commemorated by the deme of Marathon with a dedication (Paus. 1.27.10).

The ratio of cattle to sheep or goats would have varied from region to region, depending on the local environment and on both local and more distant demand. Orchomenos granted grazing rights to a man of Elateia in Phokis for 220 cattle or horses and 1000 sheep or goats (*IG* VII, 3171 (*SEG* XXVII.63), third cent. B.C.). In the fourth cent. B.C. Jason of Pherai in Thessaly made a levy for a festival of Apollo of 1000 oxen and over 10,000 sheep, goats and pigs (Xen. *Hell.* 6.4.29). This passage shows what was regarded as suitable proportions for sacrifice and also what the region could supply.

There is little detail available on the age and sex of cattle slaughtered. No calves are specified in the deme calendars. The state calendar of sacrifices (Table B) included examples of a new-born (?) *leipognomon*, costing 25 drachmai. The cattle bones at the Ephesian Artemision are described as subadult (between 1 and 3 years old?). There may have been local variations in how soon the term *teleios* ('full grown animal') could be applied. On Keos it seems to be replaced by a phrase referring to the loss of all milk teeth, i.e., aged 3-4 years (*ois* or *bous beblēkōs*, *LSCG* 98, 7 and Hansen, 1973: 329-30). In view of the modest number of cattle sacrificed (with the exceptions to be discussed below) and the importance of cattle for work, it is safe to say that the strategy generally adopted in the classical and Hellenistic periods was basically the first of the three outlined above, to produce work animals with a small surplus for slaughter. ('Male cattle are more suitable for sacrifice and work than females, but of other stock the female is more valued'. Paus. 7.22.11.) There was usually no concern with milk for human use (cf. White, 1970: 277, citing Plin. *N.H.* 25.94 for the conspicuous exception of Arkadia). However, the reluctance to sacrifice work animals meant that most of the adults sacrificed had been kept for this purpose. The preference for juveniles and subadults seen in the animal bones also points to modification of the first strategy in the direction of meat production, the third strategy.

Where extraordinary amounts of money were available for ritual and social use, a different pattern can be seen. Classical Athens is the most obvious example but there are indications of similar developments at Syracuse and, in the Hellenistic period, wherever rulers built great altars or wealthy benefactors sacrificed for the whole community. In Athens the democracy through the mechanism of the state provided the many with the style of life of the few (cf. [Xen.] *Ath. Pol.* 2.9). Cattle were the most numerous victims of the great public sacrifices of the Athenian democracy, which the exceptionally wealthy Athenian state could afford to provide. In the year 410/409, while Athens was still at war with the Peloponnesians, 5,114 drachmas, say a day's wages for 5,114 men, was spent on 100 cows for the Great (four-yearly) Panathenaia (*IG* I^3 375, 7 = ML 84). The meat was shared out among the demesmen participating in the Great Panathenaia (*IG* II^2 334) and public distribution was the rule. For the year 334/33 the income recorded for the sale of some, not all, the ox-hides from state sacrifices amounted to over 10,000 drachmai (*IG* II^2 1496 = *SIG*^{3} 1029 – receipts for only seven months are preserved, amounting to 5,099 drachmai and 4 obols. See Appendix on the value of hides and the number of cattle involved). Between 1,400 and 1,700 hides per year, I would estimate, were sold by these officials alone. Clearly cattle were important for the Athenian democracy.

Athens undoubtedly had a great many working oxen – ploughing the fields, hauling stone from local quarries and silver from local mines and imported timber and other heavy items from the harbours (but even so a thousand yoke had to be imported in 329 B.C. for a large building project, *IG* II^2 351, 18-19). As working animals they themselves were not regarded as suitable for sacrifice. Furthermore, Attica, aside from Marathon, did not have well-watered pastures for cattle.[13] We do not hear of herds of cattle. It may be significant that when Thucydides speaks of the effects of the Peloponnesian War on livestock (being sent to Euboia and the islands, 2.14, or lost to the enemy, 7.27.5), his words are *probata kai hupozugia* and he does not mention *boskēmata* ('animals reared for meat', such as Meidias procured for himself in Euboia, Dem. 21.167).[14] The greater suitability of Attica for grazing than for agriculture (Plut. *Solon* 23.4, given as explanation of the bounty on wolves) need refer only to sheep and goats.

Small-scale, intensive mixed agriculture, important in areas of high population, such as classical Attica, permitted a certain amount of livestock, more significant in toto than on any one property (cf. Hodkinson in this volume, without specific reference to cattle). In the sale of confiscated property in the late fifth cent. cattle are not conspicuous. On one property there were two work oxen, another two *boes* without qualification, and then four with calves, followed by 84 sheep and 117 goats, both herds with their young in addition.[15] The maintenance and breeding of working stock would have permitted a modest off-take of younger animals for sacrifice; the consumption of old work animals, however inappropriate for public ritual, would have provided an addition to the diet. In Greece in the present century, even in areas where cattle have been kept primarily for work, as on the island of Karpathos (personal observations), the vowing of an ox or cow to a saint for a festival is widely attested (cf. Georgoudi, 1979). Modest sacrifice, as publicly at Marathon and Thorikos, and personally by more prosperous farmers, could have been made from local cattle. However, in pre-modern conditions no herds increase rapidly (cf. Dahl and Hjort, 1976: 71-5) and one suspects that the reproduction rate of working cows is less than that of grazing herds. In Attica the local fodder would have been needed for the working and breeding animals but if grain could be imported for human consumption fodder could also be obtained, despite the expense, for fattening the cattle the population would consume.[16]

It is, however, the scale of Athenian state expenditure that argues for access to herd animals. The model of the southern Argolid, in the much more sparsely populated years around A.D. 1700, suggests that upland and borderland pastures were used as reserves for cattle, even if less desirable than lowland meadows. Evidence for Attica is lacking but the Skourta plain, shared by Boiotians and Athenians (Thuc. 5.42.1), would have been suitable. It has long been supposed that Boiotia and Euboia, with their revealing names, and Megara and the Peloponnesos were the chief sources of Athenian cattle, and it is at least a reasonable inference.[17] Large herds of up to 220 cattle or horses could be accommodated at Orchomenos with its meadows around Lake Kopais (*IG* VII. 3171 – I take the two categories permitted in that inscription, large and small stock, to show that different grazing for each was available).

In the classical period Athens may have been one of the few places where local supplies of cattle were inadequate. The Panhellenic centres, Delphi, Olympia and Epidauros, are situated in regions with relatively large areas of grazing and thinner populations, and the need for large numbers of sacrificial animals would have come at longer intervals than in Athens which was said, no doubt in exaggeration, to have had twice as many festivals as other states ([Xen.] *Ath. Pol.* 3.9). Only studies of the particular regional environments can suggest what regimes were used for the rearing of cattle. The driving of cattle over long distances to seasonal pastures is uneconomical (Cherry, in this volume), and may not have been easy in the fragmented political landscape of classical Greece, but is practised when the value of cattle is high (cf. White, 1970: 280). It is hinted at in the Homeric Hymn to Hermes where the infant god performs marvelous feats of cattle-driving in the course of cattle-rustling and plans to graze his herds on both mountain and plain (*Hom. Hymn* 4. 491-2).

Pigs

After cattle the next most expensive full-grown animals were pigs. Pregnant sows (probably only to agricultural goddesses, e.g. at Marathon, *LSCG* 20, A 43, B 48-9, costing 20 drachmai) and other adult pigs (notably five for the Attic *genos* of the Salaminioi, at 40 drachmai each, *LSCG, Suppl.* 19, 87 ff.) are prescribed in cult inscriptions (cf. the Roman *fordicidia*, bones of whose victims have been found, Ampolo, 1980: 36, n. 116). But the great majority of pig sacrifices are of young animals costing usually 3 drachmai in classical Athens. The Erchia calendar lists 10 piglets out of 56 victims, none of which are full-grown pigs. The piglets were used primarily (1) as minor offerings to figures who needed to be recognized in a larger complex of sacrifices, as at Erchia (*LSCG* 18), for the minor figure Kourotrophos at a festival of Artemis in the first month of the Attic year; (2) as the preferred victim for Demeter (cf. Jarman, 1973); and (3) as victims for purification (e.g., at the opening of the Athenian assembly; Parker, 1983: 30, 283).

While a supply of piglets was important for ritual, they offered little in the way of food, and the purification and some at least of the Demeter rites in which they were commonly used did not involve their consumption. However, a considerable number of infant pig bones were found in the central room of the Temple of Apollo at Halieis, probably for the meals of priests, though possibly for purification or divination. There was sometimes a positive preference for young animals: on Keos, if a pig is offered it should be less than 18 months old (*LSCG* 98, 8-9), but for a feast at Lindos on Rhodes a full-grown male is specified (*LSCG* 140). Aristotle mentions the frequent inability of sows to nurse all of a litter (*H.A.* 573A). The removal of a few newborn piglets from the usual litter of 7 or 8 would not affect the ultimate size of the herd. In cost, adult victims at 20-40 drachmai in fourth cent. Attica were between sheep and goats (10-15 drachmai) and cattle (mostly 50-90 drachmai). Aside from the cult of Demeter, Ge and their like, it is not clear that there were positive ritual reasons for the use of adult pigs. In some cults pigs were forbidden (cf. Kadletz, 1976: 23-4, Aphrodite, and *LSCG* 114 and 126, *LSCG, Suppl.* 73; but cf. Reese, forthcoming). Their absence at Erchia has to do with the modesty of that whole calendar and its lack of a Demeter festival, while at Marathon they appear only for Demeter. Consequently the specification of five (only one of which is for Demeter) by the Salaminioi (*LSCG, Suppl.* 19) at a cost of 200 drachmai out of a budget of a little over 500 suggests that the *genos* had access to good animals at that price.

The value of pigs, unlike that of other potential sacrificial victims, is solely for food, especially as a source of fat (cf. Sivignon, 1976 – one should not assume that olive oil was permanently and easily available). Their rearing has clearly had a long history. They are listed on Mycenaean palace tablets, in herds and as smaller numbers of fattened animals (Ventris and Chadwick, 1973: 132). The swineherd Eumaios is an important character in the *Odyssey*. In the Homeric poems pigs are grazed extensively (herds of pigs at *Il.* 11.670-84, *Od.* 14.96-104) and Eumaios and his helpers live at a distance from the town. In areas with oak trees they feed seasonally on acorns and may be herded extensively (cf. Decasos, 1904: 129). Their rooting, digging habits may be useful on fallow land (Halstead, 1981: 323) but can be dangerous to more intensive agriculture if not closely controlled. The damage done by wild boars to 'the works of men' reaches mythical proportions (e.g., *Il.* 9.538-42, Hdt. 1.36). But pigs are potentially the best suited of all sacrificial victims to intensive rearing since they thrive on household and garden waste and are easily kept penned. Aristotle mentions supplementary feeding of barley to rutting boars and to sows after they have littered (*H.A.* 573B). The figures for pigs ca 1700 A.D. in the southern Argolid suggest that, except in the most pastoral, upland village, where extensive herding is likely, they were kept, in small numbers (less than one per family), close to the fields and gardens. (See note 10, above.) Such a regime would have suited the needs of the more densely settled classical Attica. But, as with all food for that exceptionally large concentration of population, the resources of neighbouring, more rural areas were also drawn upon (cf. Aristoph. *Ach.* 764-73; the neighbouring *poleis* are the suburbs of the Athenians, Heracleides 1.2 Pfister).[18]

In sum, pigs were used in only a limited way for public sacrifice and adult animals (other than pregnant victims) may have been sacrificed mostly by private groups. But in areas of more dense population they seem to have been reared only in moderate numbers. The classical farm without the more abundant rootcrops and wastes of today may not have been as well adapted to their intensive rearing. Plato (*Rep.* II, 373C) associates them with comfortable living rather than with subsistence or piety. It is the luxurious city for which swineherds are necessary, as is also the rearing of all sorts of animals.

Sheep and goats

At every turn we have seen that in the classical period, even in Attica away from the state's largesse, the predominant victims were sheep and goats, the latter being more restricted in ritual (e.g., *LSCG, Suppl.* 73 and 74, Thasos; Paus. 2.26.9, Epidauros; sheep are prohibited for Apollo and the Nymphs on Thasos,

LSCG 114). Apollo, Artemis, and Dionysos particularly favoured goats (Richter, 1972: 426-7). The inscriptions can be quite specific about the kinds of victim to be offered. Much of this has to do with the expressive character of the victim in the symbolic system, as the occasional requirement of pregnancy or potency shows. But, to some degree, considerations of what animals were available at particular times of the year will have had a bearing.

Largely from the work of ethnographers on both transhumant and sedentary pastoralists in Greece a picture of varied and flexible strategies has emerged in recent years.[19] Climate and topography set the basic conditions in which the keeping of sheep and goats takes place but greater or lesser dependence on the results of agriculture can modify the patterns of adaptation to natural conditions. An extensive system makes more use of natural pastures and less of the incidental or intentional products of agriculture (e.g., weeds in an orchard or fields planted in legumes for grazing). An intensive system involves more labour and higher yields per animal. The symbiosis of pastoralism and agriculture is characteristic of all lowland areas of Greece, though not limited to them, in all historical periods, but it takes various forms, and the relative importance of the products of each does not remain fixed.

As a particular example of the combination of practices to achieve more than one goal, let us consider the pattern that has prevailed in Greece in recent times. Herding has been geared to a regime in which the production of milk, converted primarily into cheese, is combined with the sale of animals for meat, primarily young males born in the winter and spring, and of older animals no longer useful for breeding, milking or wool. The availability of natural pasture has established a cycle of breeding in the summer and birth in the winter and the spring – not, to be sure, universal, so that there are in fact births at other times. The main festivals of the church, at Christmas and Easter, and the demand for meat at Carnival, suit the removal of young animals over and above those needed to maintain or, if desired, to increase the flock in winter and spring. Unwanted adult animals are usually culled from the flock in the summer but when the shortage of summer pastures was more severe before the advent of plentiful supplementary feed and when the movement of flocks to higher pastures was more difficult, it is likely that the extra animals were slaughtered before the full strain of summer was felt. Today a settled herder's flock in a good year may double in size with the birth of the new lambs and kids because most of the animals are females, kept for milk and breeding, rather than the larger males with more meat and, in the case of sheep, better wool. However, it is only recently with the use of medicines, the elimination of wolves from most of Greece, supplementary feeds and field crops for fodder facilitated by increased irrigation that only a small percentage of the animals are lost each year. Earlier less than half of the new animals – and young females only with great reluctance – could be removed if the size of the herd was to be maintained.

In the deme calendars from Attica, Marathon offered no young sheep or goats, Thorikos 6 young animals out of 35 sheep or goats (1:6) and Erchia 5 young sheep or goats out of 36 (1:7).[20] These are all community festivals in which numbers of people participated and were fed and we should expect, therefore, more adult animals than in smaller group or family sacrifices. The male sheep, larger than the ewes, were probably castrated wethers since only a single ram is specified at Erchia and again at Marathon. The greater number of young sheep and goats at Erchia along with the absence of cattle seems to be an economy measure. Most private persons would have offered more sheep and goats than cattle and more small than large animals (cf. Hdt. 1.133.1). That the choice between older and younger animals was the main decision the private sacrificer faced is seen in a number of inscriptions where different fees are prescribed according to whether a *teleios* ('full-grown') or a *galathēnos* ('suckling') has been sacrificed, e.g., *LSCG* 45,4-5 (cf. Stengel, 1879: 7-11; Ziehen, 1939: 595-7). Where cult officials took their shares from the animal sacrificed or a group expected to dine on the meat there was practical interest in the size of the animal and the quantity of meat available. Larger rather than more numerous animals tend to be prescribed in such situations.

Of adults Marathon sacrifices males and females in almost equal numbers, but at Erchia the ratio is 2:1 and at Thorikos only a little less than that. (For lambs and kids the numbers are too few for the ratios to be useful.) Cult requirements seem to permit a preference for slaughtering males at Erchia and Thorikos. For effective herd management today, with the emphasis on milk production and secondarily on meat, herdsmen cut out not only the majority of lambs and kids before they are a year old but also ewes and does past lactation (at six years) and rams and bucks from three or four to eight years (Harold Koster, personal communication). But from the calendars we do not know how the adult victims are distributed in age as between milk and wool animals past their prime and young adults raised for sacrifice, for whom the optimum age in terms of the amount of meat gained from the amount of feed consumed is between two and three years (Payne, 1973: 281, who warns [personal communication] that this optimum applies only to a fairly extensive system with no marked seasonal shortage; in modern Greece less than one year old is optimal).

The archaeological evidence is mixed. At the Ephesian Artemision the sheep and goats are largely female and older than a year but it is not clear how much older. At the Theban Kabeirion juveniles and subadults predominate, but whereas most of the goats are young, the sheep are older, perhaps because of the value of wool production (Boessneck, 1973: 14; no more precise description of ages is given). At Halieis, inside the temple, most bones are infant sheep and goat tibiae, along with infant pig tibiae. Goats older than two years, however, furnished the horns that were found in the middle room, perhaps from animals sacrificed at the alter before the temple.[21] The youngest animals may be underpresented in the published evidence because of the fragility of their remains in general and inattention to a useful criterion for age (Payne, 1985: 230). Overall, there are two main groups – infants and young adults – the former may represent the culls consistent with a milk strategy, the latter with one aimed at meat. The bones, insofar as they are relevant, do not disagree with the calendars and suggest that young adults (one to three years old) satisfied the inscriptions' specification *ois* or *aix* with or without the adjective *teleios*. The ratio of young to adult victims (1:6 at Erchia, 1:7 at Thorikos) does not suggest as heavy a rate of reproduction and of slaughter of the very young as might be expected. If many of the adults are in the young adult or subadult category, we would have a modification of a regime geared to milk and wool in the direction of meat production.[22]

The notion of not weakening the herd by killing off replacement stock, primarily young females, is explicit in an *aition* given by the Atthidographers to explain why the priestess of Athena Polias does not sacrifice a ewe lamb (*amne*) or taste cheese (Androtion *FGrHist* 324 F55 ap. Athen. IX. 375B-C): 'There was an old law for the sake of the reproduction of young not to slaughter a ewe which has not been shorn of wool and has not given birth.' In the version attributed to Philochoros (*FGrHist* 328 F169 ap. Athen. I-9C-D) the prohibition is extended to all Athenians, which cannot be right since Athena on the local *polis* hill at Erchia received a ewe lamb, and an historical shortage is given as the cause of the concern. The combination of the prohibited victim with the priestess's avoidance of cheese (fresh cheese and Attic cheese but not foreign cheese, Strabo 9.1.11) shows that we have to do with a complex. symbolic expression. The explanation given is an indication of the fourth-century writers' interest in agriculture and husbandry (Androtion was the author of a *Georgika*, *FGrHist* 324 T17) but if more than fragments of the Athenians' ritual and symbolic code had survived, we might be able to see the expression of an older and broader concern with the preservation of their resources. Pragmatism and hedonism are combined in a comic reference to 'putting up' with the inferior (*phaula*) meat of lambs and kids for the sake of the profits from the cheese and wool of full-grown animals (Antiphanes II 17 Kock, ap. Athen. IX. 402D-E).

The demands of any one purchaser in a country district for between 50 and 60 animals in the course of a year, which were the needs of the three demes, will not by itself disclose the local pattern of herd management. But if, as seems likely, they represent the pattern of religious observances that extended throughout the community their implications for the pastoral calendar are worth examining.

Generally, the young animals are specified at times of the year when they would be plentiful.[23] At Erchia the young sheep and the kid are sacrificed in February, March, July, September and October, in all of which young animals would be easily available according to the cycle we can observe today. The periods of greatest sacrificial activity are February, May-July and September-October. In February (Gamelion), aside from two piglets, nine full-grown animals and one ewe lamb are sacrificed, and all but one of the adults are male. A season when all attention and food needed to be devoted to the nursing mothers and their young was a good time to slaughter extra males. In May-July, 11 adult males, two ewes and one doe and one male lamb were sacrificed. This is the best period for culling a flock before turning it on to the scarcer summer grazing. The lamb is for Zeus on the Heights (Epakrios) and is to be sacrificed and consumed on Mt. Hymettos – a fee paid, so to speak, for the mountain pastures the animals have begun to use by mid-June.[24]

At Thorikos in south Attica sacrificial activity is concentrated between March and July with the peak coming in May, and five of the six young victims (other than pigs) are sacrificed between March and July, as are two pregnant ewes. These last may not be as wasteful as might first appear since they are killed before the onset of a period of poor grazing for flocks not moving to high country.[25] There is another spurt in October and November when the stubble in the lowlands has been used up and the new grass is yet to come. The first two months of the Attic year are lost for Marathon and there are adjustments between the deme and the Tetrapolis that obscure the picture. Spring is busy but with six cattle, two sows and no lambs or kids offered this calendar has less bearing on local pastoralism.

On the basis of these Attic calendars it may be suggested that in most of Greece, without the Athenian state's resources to buy many animals for sacrifice, the sacrificial calendar was close to that of seasonal availability from the annual increase of young animals and the culling of older ones, and that in Attica as well as elsewhere the production of milk and wool was the prime goal of the owners of flocks of sheep or goats.

Milk production, chiefly in the form of cheese which is more easily preserved than meat, makes a far more effective contribution to the diet than killing animals for meat. Greek goats today 'give more than six times as many calories in milk as in meat, and nearly three times as much protein' (Payne, 1985: 226). We should not suppose the people engaged in the exhausting work of herding were not aware of the results of their labour, to the extent that they were free to make choices. Hired and slave herdsmen, of course, would pursue whatever policy the owners found profitable or socially desirable. The existence of the large urban population of Athens and the purchasing needs of the Athenian state are likely to have altered the regimes of what had been relatively self-contained communities, as has been the case in modern times.

Sources of sheep and goats

In Attica the demands for sheep and goats comes primarily from smaller groups such as the demes, *gene* and families. It is not evident that their needs could not have been satisfied by local Attic pastoralism, though no doubt sheep and goats, as well as pigs and, as we have supposed, cattle, were brought to market from neighbouring areas. When the Athenians sought to fulfill their vow to Artemis to sacrifice a doe (she-goat) for every Persian slain at Marathon (6,400, Herodotus 6.117.1) not enough goats could be found and they settled on 500, which became the number from then on (Xen. *Anab.* 3.2.12).[26] Since over 6,000 does would have been needed in addition to whatever normal sacrificial obligations were in effect, the removal of that many breeding animals in the autumn (Boedromion) and without planning if soon after the battle, could well have disrupted local herds with the small surpluses available for slaughter under pre-modern conditions. The story does not really reflect on the adequacy of the local goat supply.

There are few references in Greece proper to 'sacred herds,' that is, animals belonging to a sanctuary.[27] At Tegea in Arkadia in the early fourth century there were rules at the sanctuary of Athena Alea for the pasturing of *hiera probata*, evidently herded extensively and passing through the goddess's land. They included larger and smaller stock (cattle as well as sheep or goats) while pigs are mentioned separately. These may have been herds belonging to a god (i.e., to a sanctuary) or herds designated at some point as reserved for a particular festival. The Phokians said that whatever *boskēmata* they named as sacred to Artemis were free of disease and grew fatter than others they reared (Paus. 10.35.7). However, when Xenophon established a cult to Ephesian Artemis on his estate near Olympia, food produced and animals reared on the estate were to be used for the annual sacrifice, but at other times they were at the disposal of the lessee of the estate (*Anab.* 5.3.9-13). Unlike cattle and pigs, sheep and goats belonging to a sanctuary could produce income from milk, wool, hair and manure. But most victims did not come from sanctuary herds since the assumption in the listing of prices in *leges sacrae* is that animals will be purchased. At times animals seem to have been bought in advance and fattened up for sacrifice (Aristoph. *Eq.* 1135-6 with Schol.; cf. Stob. *Flor.* 79.8; a single special victim at Magnesia on the Maiandros, *LSAM* 32). But at Athens the scale of sacrifice makes it unlikely that this could have been done with more than a small fraction.

The evidence for Athenian pastoralism is discussed by Hodkinson in this volume, who has noted that while Athens does not include rights to grazing in its grants to foreigners, the practice is in any case limited in its occurrence in Greece. In this populous city-state, however, one would expect that most grazing was fully occupied by local herd-owners and that the issue did not arise. To Hodkinson's description I add only that a close connection with agriculture makes it easier to rear animals for a particular time and purpose. Athenian sacrificial practice with its need for group feasting and less stress on the new-born animals than is the case today, even making allowance for the public character of our sources, would be well served by a certain amount of intensification such as seems likely in the Attic countryside of the fifth and fourth centuries.

Sacrifice and diet

Important as the social and religious functions of sacrifice were in Greek life we should not over-estimate the amount of meat the ritual added to the diet. Certainly it was a welcome supplement to the blandness of cereals and legumes even when enlivened with the *opson* (olives, cheese, onions and greens) Plato concedes to the simple city (*Rep.* II.372C). But the actual weight of meat consumed and the proteins and calories it contributed were modest. With the figures for usable meat given above in the discussion of cattle (cf. Dahl and Hjort, 1976: 203; Sloan and Duncan, 1978: 76) the total amount of meat produced by the sacrifices in the Erchia calendar was approximately 796 kg.[28] Any calculation of the citizen population of the deme will be very approximate but adequate to make our point. If we suppose that the seven members of the Council of 500 from Erchia were roughly proportionate to the population of the deme as a fraction of the total population of Athens, the Athenian citizens in the deme were 1.4% of all citizens, or 1,400 in the fourth cent. when the total was around 100,000 (Traill, 1975: 67; Gomme, 1933: 29).[29] This one calendar (with two victims missing at the end) would have supplied 0.5686 kg. of meat per person. Let us further suppose that the missing 'Lesser Demarchia' supplied as much again, and that the sacrifices of other organizations and private sacrifices of the demesman did the same, so that we triple this figure to 1.71 kg. per person. The paltriness of the result can be seen when it is compared with the 11.06 kg. of meat eaten by Cretans in the year 1947, which, together with poultry, game, snails and fish, still amounted to only 30% of the per capita quantity consumed in the United States at that time (Allbaugh, 1953: 110; 7.5 ounces a week = 24.375 lbs. a year = 11.06 kg.). In effect, the Erchia Calendar would have had to be multiplied 19 times to provide a quantity of meat for the deme members comparable to that eaten by post-war Cretans. With such extreme discrepancies we can, I suggest, throw out our hazardous calculations and retain only the conviction that the local ritual alone played little part in the nutrition of the demesmen, and since the majority of Greek communities were closer, in the scale of their sacrifices, to Erchia than to the city of Athens, the general impression that the Greek diet was based largely on plant and milk products need not be modified (cf. Bruns, 1970: 61-3[30]).

In the light of these conditions the prodigality of the Athenian state can be better appreciated. The heavy expenditure for cattle produced a vastly greater quantity of meat than the sheep and goat sacrifices of the demes and most private persons. If we take 2,000 head annually as a minimum number (we estimate between 1,400 and 1,700 from the sales of hides; see Appendix) at 100 kg. of meat per animal there was 200,000 kg. available, or 2 kg. a year for each Athenian citizen (not counting foreigners and slaves). This more than doubles what we have suggested each person at Erchia might have received, and it is probably too cautious a figure. But, as has become all too clear with the rise of meat consumption in third world countries, feeding and slaughtering animals for meat is far less economical than growing cereals for food or for the fodder of milk animals. For Athens it was a social and political choice which, when she was prosperous, she could afford. In the Hellenistic and Roman periods Greek cities, Athens included, were, I believe, dependent in this respect as in others on the generosity of rich patrons, but it is a subject that remains to be examined in detail.[31]

Ritual and practical aims

Local, ritual calendars correspond roughly to the seasonal supply of various animals. There are, however, sacrifices prescribed which are strictly 'impractical' in that they have powerful religious functions but if widespread throughout the community would weaken the strength of the herd for survival and for fulfilling the other goals of the owners. These are the slaughter of pregnant victims – cows and sows at Marathon, ewes at Thorikos and Erchia, although the latter may have some practical grounds as well. So too the occasional prescription of a ram or buck, as opposed to a male sheep, i.e. wether (*ois*) or goat (*aix*), requires the keeping of a more troublesome, uncastrated male of less value for meat and wool. In small numbers these, and the even less common and more troublesome *tauros*, could be supplied by the culling of older stud animals though they might not pass the test implied by the term *kritos* (in the Thorikos calendar, cf. *kallisteuōn*, e.g. *LSCG* 96.6). Peleus's vow of fifty uncastrated sheep in addition to a hekatomb to the river Spercheios (*Il.* 23.147) was meant, no doubt, to be on a heroic scale. Nonetheless it is inescapable that, while sacrificial practice seems generally consistent with the other reasons for maintaining animals, the primarily expressive character of certain rules has purposes of its own.

An example of this is the prohibition of goats as victims for Asklepios at his great sanctuary at Epidauros (Paus. 2.26.9, cf. 10.32.12, and perhaps at Erythrai, *LSAM* 24.8-10). The mountainous Argolic Akte is particularly well suited to goats. Their grazing was probably the chief issue in the long drawn out border dispute between Epidauros and Hermion and is likely to have been at the heart of similar disputes between Epidauros and Corinth and Epidauros and Arsinoe (Methana).[32] Goats might be sacrificed to all the other gods of Epidauros, particularly perhaps to the earlier chief god of the area, Apollo, but not at the sanctuary of the healing god. One might see in this a desire on the part of local cult officials to encourage visitors from abroad to make more expensive sacrifices by which the god would be more honoured and from which the officials would get more and better meat. Such a jaundiced view may not be justified. Ritual at times seems perverse and even harmful economically, though we cannot say this was actually the case in the Epidauria, but it serves equally important social purposes. At Epidauros the rule may have underlined a distinction between Apollo with his occasional preference for goats and the newer patron of the sanctuary, Asklepios, whose Panhellenic status was reinforced by his disdain for the common local beast. Furthermore, if the principal aim of the local sheep and goat herders was the production of milk, with their surplus going to honour the numerous other local cults, they may not have been much affected by the rule in the Asklepios sanctuary.

An anthropologist who has given much thought to the relationship of subsistence and ritual has coined the phrase 'a ritually regulated ecosystem' (Rappaport, 1979: 41). The closer a society is to having a self-contained, subsistence economy the better this concept applies. When wealth is concentrated with less direct relation to local agricultural or pastoral resources, then social and ritual values become more independent and sacrifice becomes part of what might be described rather as an economically limited ritual system. Such a situation was coming about in Classical Athens and in time it probably characterised much of the Greek and Roman world, which would have approved the priority given to sacrifice by the pious gentleman farmer: 'The art of rearing animals is joined to agriculture so that we may have the means to please the gods by sacrificing and may make use of them ourselves' (Xen. *Oec.* 5.3).

This paper was written while I was a fellow of the Center for Advanced Study in the Behavioral Sciences, Stanford, California. I am grateful for financial support provided by the National Endowment for the Humanities #FC-20029-82 and the Andrew W. Mellon Foundation.

STANFORD UNIVERSITY MICHAEL H. JAMESON

Appendix *The Value of Ox-hides*

It has been asserted since August Böckh (1886: I, 97 and II, 22*) that three-drachmai and four-and-a-half drachmai prices for an ox-hide are attested in fourth-century Athens (Böckh was followed by Ziehen, 1899, and Ferguson, 1944: 101). Neither of the two passages cited by Böckh permits this inference.

The first passage is the cost per item of 17 *diphtherai* for 17 public slaves at the Eleusinian sanctuary (*IG* II^2 1672, 104, 329/28 B.C.). This entry is preceded by another for items of clothing. The *diphthera* is likely to have been a tunic made from goatskin (cf. Aristoph. *Nub.* 72).

The second passage cited by Böckh comes from an inscription (*IG* II^2 1356 = *LSCG* 28) which lists the payments to be made to priests in kind or in cash for a series of sacrifices. Normally the priest receives the skin of the animal sacrificed.

For sacrifices of victims that are known as *heusta* ('singed'), which do not have flayable skins, *darta*, a cash payment of three drachmai is prescribed. For the meaning of *heuston* see Ziehen, 1899: 267-74 and 1906, p. 81; cf. *LSAM* 36,5 (*SIG*³ 1037), from Miletos, and *LSCG, Suppl.* 19, 32-3 (*Hesperia* 7 (1938) 1-68) from Athens. For *darton*, Stengel, 1910: 131, n. 3. But there was no reason why compensation for the priest who sacrifices a victim without a skin convertible to cash (i.e., a pig of whatever size) would have been equivalent to the full value of an oxhide from the largest and most expensive victim. In fact, *IG* I³ 255 **B**, 14-18 (= I² 190 = *LSCG* 11) and *LSCG, Suppl.* 19, 32-3, show that while the skin of a flayable victim might be included in a priest's perquisites, for a singed victim (*mē darton* in *IG* I³ 255) he could have his normal perquisites but no compensation for the lack of a skin. At Delphi sacrificers from Skiathos could take the skin if they paid the sanctuary two obols for a public sacrifice, one obol for a private sacrifice (*LSCG, Suppl.* 41, 12-15). The conclusion must be that the three drachmai of *IG* II² 1356 was not the equivalent of the skin of any particular type of victim (which might have been large or small, male or female, and cattle, sheep, or goat) but an arbitrarily fixed fee to be paid to the priest. This suggests that a hide from the most expensive sacrifice was almost certainly worth more than three drachmai.

Despite the heavy slaughter of cattle in fourth-century Athens the demand for hides remained high and it was profitable to import them in large quantities. A ship's captain overloaded his ship in the Pontos with 1,000 *dermata* (presumably ox-hides), thereby, it was alleged, wrecking the ship ([Dem.] 34.10; goat skins were also purchased in those parts [Dem.] 35.34). In the fifth century ox-hides were also imported from Kyrene (Hermippus 243 Kock, ap. Athen. I, 27E).

The interest in arriving at an approximate price for an ox-hide is in applying it to the receipts recorded for the sale of hides by the state in 335/34 B.C. and thus obtaining a minimum figure for the number of cattle sacrificed in that year. But it is easier to show what a hide was not worth than to say what it was. The following is a frankly speculative effort to do so by estimating the relative value of hides to wheat in antiquity. An unworked ox-hide of first quality has its price fixed at 500 *denarii* in Diocletian's Price Edict (Lauffer 1971: **8** 6a, pp. 126-7). Unfortunately no prices are preserved from the Edict for cattle. The price fixed for wheat is 100 *denarii* for a *modius castrensis* (**1** 1a, pp. 98-99), which has been shown to be equivalent to $1\frac{1}{2}$ times a *modius Italicus*, 12.315 litres and 8.6185 litres, respectively (Duncan-Jones, 1976a: 43-52 and 1976b: 53-62). The Attic *medimnos* may be taken to hold 52.176 litres (Lang and Crosby, 1964: 46) and since it was therefore about 4.27 times as large as the *modius castrensis*, that amount of wheat would have been priced at 427 *denarii* in the Edict. A hide, at 500 *denarii*, was valued at 1.17 times the cost of a *medimnos* of wheat.

To apply the information from the Edict to Classical Athens requires our making two assumptions: (1) that the relation between the price of cattle and that of wheat was much the same in the two periods, and (2) that an ox-hide was valued at approximately the same fraction of the cost of the ox in the two periods. Examination of the first of these assumptions involves something of an excursus.

The evidence of prices from Medieval and Early Modern western Europe shows a close relation between the price of wheat and that of cattle (Slicher van Bath, 1963: III B, especially 98 and 115). In antiquity wheat and ox-hides, universally available and not affected by workmanship applied to them, may also be expected to have had a fairly stable relationship in price. At any particular time the prices of cattle and wheat, of course, varied and would diverge to some degree from their average corrrelation over the long term. Diocletian's Edict was, notoriously, not concerned with actual price fluctuations or supply and demand. For sacrificial animals in Classical Greece the price may be expected to have been more closely tied to grain prices than that of other livestock since they were reared solely for butchery and would have been fattened on fodder as much as possible (cf. Columella 2.9.14, wheat as fodder). For Classical Athens differences in the sources of supply had a bearing. About half of Athens' grain came from overseas (Garnsey, 1985), whereas cattle (unlike ox-hides) are not known to have come from beyond Attica itself and, probably, nearby areas in east central Greece. Crop-failure in this region could be more easily countered by increased imports than could comparable shortages of cattle but the latter would not develop as quickly.

In Athens in 414 B.C. we have prices for both wheat and cattle from the same document, the forced sale of the property of those involved in the scandals of 415 (Pritchett, 1956. 196-8, 257: ML 79, 68-70, who suggest all prices of these forced sales may be on the low side). Wheat sold for between 6 and $6\frac{1}{2}$ drachmai and cattle for 35 and 50 drachmai each. The ratio is 1:5.3-8.3 but the essential figures are probably $6\frac{1}{2}$ for wheat (Pritchett suggests the 6-drachmai wheat was of poorer quality) and 50 for cattle, if the two 35-drachmai animals, evidently inferior to the two 50-drachmai plough-oxen in some respect, are left out. (That the price of cattle in the fifth century was generally in this range can be seen from Thorikos (40-50 drachmai) and the animals for the Panathenaia of 410/9, 51 drachmai each (ML 84, 7 = *IG* I^3 375 (I^2 304A).) In this case the ratio is 1:7.7. At this time an ox-hide, if it was 1.17 times the price of a *medimnos* of wheat, as in the Edict, would have cost approximately 7.6 drachmai and its relationship to the price of an ox would have been 1:6.6.

I know of no other example of clearly comparable wheat and cattle prices. On Delos from the end of the fourth to the mid-second century there was generally a 1:7 ratio with wheat around 10 drachmai and cattle at 70. (When on occasion cattle rose to 120 no wheat prices are preserved, except possibly in 190 B.C., at 10 drachmai; Heichelheim, 1930: Table XIV.)

The fourth century presents particular problems. For the first half of the century, when we have the high cattle prices, by late fifth century standards, of 100 drachmai (Panathenaic prizes, perhaps exceptional animals, *IG* II^2 2311, 72-

6), 90 drachmai (Marathon), 77 (375/4 B.C., for Delos, *IG* II[2] 1635, 35-6) and 70 (363/2 B.C., Salaminioi, *LSCG, Suppl.* 19, 85), I know of no prices for wheat. (The figure of three drachmai cited from Aristoph. *Eccl.* 547-8 in 393 B.C. should not be pressed. Blepyros exclaims that he has lost a whole *hekteus* of wheat (a sixth of a *medimnos*) by not going to the assembly, where we are to understand he would have picked up three obols of pay.) In the second half of the century, 5 drachmai was regarded as the normal price for wheat and prices of 5-6 are attested between 330 and 324, with an exceptional 9, once, in the period 340-30 (Pritchett, 1956: 197-8). The figure of 4,100 drachmai in 335/34-330/29 (*IG* II[2] 234, 16 = *LSCG* 33) comes from income of rented land and has been thought to be for less than a hekatomb (Ziehen, 1896: 215). But at a time when wheat was 5-6 drachmai, a cattle price of 41 drachmai and a ratio of 1:6.8-8.2 would not be out of line. If a standard relationship existed, as I believe it did, I suggest it was close to 1:7-8.

Against this must be set the astonishing price of 400 drachmai for cattle and 30 for sheep and a goat in 329/8, a vivid illustration of religious considerations being paramount at a time of drought and short supply (*IG* II[3] 1672, 289-90; Hansen, 1975: 200-8, who suggests that the two preceding years had been more severe). But in the same year as the Eleusinian officials purchased these extraordinarily expensive cattle they sold a little over 80 *medimnoi* of wheat received as first-fruits by the sanctuary for 6 drachmai a *medimnos*. The price was set by the *demos*, as was also the price to be paid for the sacrificial victims. The distortion of normal price relations is shown by the 1:13 ratio of sheep/goats to cattle, in contrast to the usual 1:5. Clearly grain was considerably higher than the price the assembly set for what the city had available to sell. But the crisis in grain seems to have eased this year and we may doubt that it approached the 50-drachmai price of a ratio of ca 1:7.7 between wheat and cattle, suggested by the evidence of 414 B.C. The effect on cattle prices of climate problems would have been slower to appear but also slower to remedy. It is notable that sheep and goats, which we have supposed could have been pastured in adequate numbers in Attica for normal needs, did not rise as steeply as cattle.

The prices reported from Solonian legislation of the early sixth century equate a *medimnos* of wheat with a sheep and show cattle as five times as expensive as sheep (approximately the relative weight of usable meat), for a wheat:cattle ratio of 1:5 (Plut. *Solon* 23.3; Plutarch does not specify the grain but with this ratio it cannot have been for anything less valuable than wheat). If reliable, these figures would indicate that livestock were less expensive in the early sixth century than in the much more populous Classical period, when sheep were never less than 10 drachmai while wheat cost 5-6.5 drachmai and cattle remained approximately five times as expensive as sheep (cf. Böckh, 1886: I, 97). Reduction of grazing land with the expansion of agriculture and consequently a relative rise in the cost of livestock to cereals is to be expected.

It would be of considerable interest if a further rise in the relative price of cattle to grain could be demonstrated for the fourth century. Meanwhile it is not unreasonable to use the late fifth-century ratio of 1:7.7. As to whether that approximated the ratio in the Edict, we can only appeal to general considerations. Beef was less expensive than pork and the same price as mutton and goat meat in the Edict (**4** 1a-3) which argues for a relatively low price for cattle (cf. Pritchett, 1956: 257-8, who supposed beef was cheap in classical times as well). This does not suggest cattle prices were, relatively, higher than in the fifth and fourth centuries B.C.

As for the second assumption we have to make, that a hide had relatively the same proportion of the total value of the animal in all periods, there may be some indication that the 1:6.6 ratio suggested above for 414 B.C. is in the right range for ca A.D. 300. An Italic pound of beef (0.32745 kg.) was to sell for 8 *denarii*, according to the Edict (**4** 2). If 100 kg. of dressed meat was available from each animal, that comes to 305 Italic pounds, worth 2,443 *denarii* (without the higher prices for special parts such as liver). Adding the value of the hide, 500 *denarii*, the whole animal produces close to 3,000 *denarii* or six times the value of the hide. One consideration would raise the relative value of the hide somewhat: since some figure for the expense of butchering would have been reckoned into the price of meat, the cost of the live animal should have been less than 3,000 *denarii*, and the hide, therefore, worth more than one sixth of its cost. But another consideration lowers its relative value: I suspect that cattle of the third century A.D. may have been larger than the 200 kg. animals assumed for Greece in the Bronze and Iron Ages and that more than 305 Italian pounds of meat, some of which would have sold for more than 8 *denarii*, could have been taken from their carcases. (Cf. Bökönyi, 1974: 128, on the greater size of improved Roman cattle as compared to local Hungarian animals). The ratio of 1:6.6 suggested for the late fifth cent. does not seem far off the mark.

Returning to the relationship of wheat to hide, if we use the ratio in Edict of 1:1.7 and if the relationship between wheat and cattle remained what it was in the late fifth century (1:6.6), I suggest that when wheat sold for between 5 and 6.5 drachmai an ox-hide would have been worth roughly between 6 and 8 drachmai (which is double the fee, or more, given a priest who sacrificed a *heuston*.) The lower range of 5 drachmai for wheat was standard in the later fourth century and hides would have cost between 6 and 7 drachmai. Applying this to the receipts for the sale of skins in Athens, according to which some 10,000 drachmai would have been received in the whole year 334/33 (a little over 5,099 drachmai in 7 months, *IG* II^2 1496 = *SIG*^{3} 1029), we get hides of between 1,400 and 1,700 full-grown cattle sold by the officials. An indeterminable number of sacrifices did not come under their purview, and some hides may have continued to be handed over directly to the priests in charge, as was common when the state was not directly involved. 1,400 to 1,700 would be the lower limit of the number of cattle provided by the state for sacrifice. If, however, cattle prices rose vis-à-vis wheat, so did the price of hides and fewer cattle are represented by the income from their sale.

Bibliography

Allbaugh, Leland G. 1953 *Crete. A case study in an underdeveloped area.*

Amandry, Pierre 1984 'Os et coquilles', *L'Antre Corycien* II. *Bulletin de Correspondance Hellénique*, Supplément IX: 347-80

Ampolo, Carmine 1980 'Le condizioni materiali della produzione. Agricoltura e paessagio agrario', *Dialoghi di Archeologia* n.s. 1: 15-46

Bammer, Anton 1978 'Zur Datierung und Deutung der Bauten,' in A. Bammer, F. Brein and P. Wolff, 'Das Tieropfer am Artemisaltar von Ephesos,' in S. Sahin, E. Schwertheim and J. Wagner *Studien zur Religion und Kultur Kleinasiens. Festchrift für Friedrich Karl Dörner* (1978), I, 138-57

Böckh, August 1896 3rd ed. by Max Fränkel *Die Staatshaushaltung der Athener*

Boessneck, Joachim 1973 *Die Tierknochenfunde aus dem Kabirenheiligtum bei Theben (Böotien)*

Bökönyi, S. 1974. *History of domestic animals in Central and Eastern Europe*

Braudel, Fernand 1973 *Capitalism and material life 1400-1800*

Brumfield, Allaire Chandor 1981 *The Attic festivals of Demeter and their relation to the agricultural year*

Bruns, Gerda 1970 *Küchenwesen und Mahlzeiten (Archaeologia Homerica*, II, Q) Göttingen

Campbell, John 1964 *Honour, family and patronage. A study of institutions and moral values in a Greek mountain community*

Dahl, Gudrun and Anders Hjort 1976 *Having herds. Pastoral herd growth and household economy* (*Stockholm Studies in Social Anthropology* 2)

Daux, Georges 1963 'La grande demarchie d'Erchia', *Bulletin de Correspondance Hellénique* 87: 603-34

1983 'Le calendrier de Thorikos au musée J. Paul Getty', *L'Antiquité Classique* 52: 150-74 = 'Sacrifices à Thorikos' *J. Paul Getty Museum Journal* 12 (1984) 45-52.

Day, Leslie Preston 1984 'Dog burials in the Greek world', *American Journal of Archaeology*: 21-32

Decasos, Panajotis A. 1910 *Die Landwirtschaft im heutigen Griechenland* (Doctoral Dissertation, Leipzig)

Dimen, Muriel and Ernestine Friedl, 1976 *Regional variation in modern Greece and Cyprus: toward a perspective on the ethnography of Greece* (*Annals of the New York Academy of Sciences*, 268)

Dow, S. 1965 'The greater *demarkhia* of Erkhia', *Bulletin de Correspondance Hellénique* 89: 180-210

Duncan-Jones, R. 1976a 'The Choenix, the Artaba and the Modius', *Zeitschrift für Papyrologie und Epigraphik* 21: 43-52

1976b 'The Size of the Modius Castrensis', *ZPE* 21: 53-62

Durand, Jean-Louis 1986 *Sacrifice et labour en Grèce ancienne*

Ferguson, William Scott 1944 'The Attic Orgeones', *Harvard Theological Review* 37: 62-140

Foster, Giraud 1984 'The bones from the altar west of the Painted Stoa', *Hesperia* 53: 73-82

Frazer, J.G. 1965 *Pausanias's description of Greece*

Garnsey, Peter 1985 'Grain for Athens', in P.A. Cartledge and F.D. Harvey, edd., *Crux. Essays presented to G.E.M. de Ste Croix on his 75th Birthday* (*History of Political Thought* 6, 1/2) 62-75

Georgoudi, Stella 1974 'Quelques problèmes de la transhumance dans la Grèce ancienne', *Revue des Études Grecques* 87: 155-85

1979 'L'égorgement sanctifié en Grèce moderne', in Marcel Detienne and Jean-Pierre Vernant, edd., *La Cuisine du sacrifice en pays grecque* 271-307

Gomme, A.W. 1933 *The population of Athens in the fifth and fourth centuries B.C.*

Halstead, Paul 1981 'Counting sheep in neolithic and bronze age Greece', in Ian Hodder, Glynn Isaac and Norman Hammond *Patterns of the past. Studies in honour of David Clarke* 307-39

Hansen, Hardy 1973 'The Meaning of λειπογνώμων', *Greek, Roman and Byzantine Studies* 14: 325-32
Hansen, Mogens Herman 1975 in Signe Isager and M.H. Hansen *Aspects of Athenian society in the fourth century B.C.* (*Odense University Classical Studies*, 5)
Hansen, Mogens Herman 1986 *Demography and Democracy*
Heichelheim, Fr. M. 1930 *Wirtschaftliche Schwankungen der Zeit von Alexander bis Augustus*
Højlund, Flemming 1981 'The deposit of sacrificed animals at the entrance to the tomb chamber', in *The Maussolleion at Halikarnassos. Reports of the Danish Archaeological Expedition to Bodrum* 1. *The Sacrificial Deposit* (Jutland Archaeological Society Publications, XV:1) 53-89
1983 'The Maussolleion Sacrifice', *American Journal of Archaeology* 87: 145-52.
Jameson, Michael H. 1973-74 'Excavations at Halieis (Porto Cheli) 1973', 'Αρχαιολογικὸν Δελτίον, Χρονικά 29: 261-8
1974a 'A Treasury of Athena in the Argolid', in D.W. Bradeen and M.F. McGregor (edd.) *PHOROS*: 67-75
1974b 'The Excavation of a Drowned Greek Temple', *Scientific American* 231, 4: 110-19
Jarman, M.R. 1973 'Preliminary Report on the Animal Bones', in J.N. Coldstream *Knossos. The sanctuary of Demeter* (British School of Archaeology at Athens, Supplementary Vol. 8) 177-79
Kadletz, Edward 1976 *Animal sacrifice in Greek and Roman religion* (Dissertation, Department of Classics, University of Washington)
Kavadias, Georges B. 1965 *Pasteurs-nomades mediterranéens. Les Saracatsans de Grèce*
Kayser, B. and B. Thompson 1964 *Economic and social atlas of Greece*
Koster, Harold A. and Joan Bouza Koster 1976 'Competition or symbiosis?: pastoral adaptive strategies in the Southern Argolid', in Dimen and Friedl, 1976: 275-85
Koster, Harold A. 1977 *The ecology of pastoralism in relation to changing patterns of land use in the Northeast Peloponnese* (Dissertation, Department of Anthropology, University of Pennsylvania)
Lang, M.L. and M. Crosby 1964 *The Athenian agora, X. Weights, measures and tokens*
Lauffer, Siegfried 1971 *Diokletians Preisedikt*
Loukopoulos, D. 1930 Ποιμενικὰ τῆς Ρούμελης
Marinatos, Sp. 1936 'Le temple géometrique de Dréros', *Bulletin de Correspondance Hellénique* 60: 214-85
McDonald, William A., William D.E. Coulson, and John Rosser 1983 *Excavations at Nichoria in Southwest Greece.* Volume III. *Dark age and Byzantine occupation*
Meiggs, Russell and David Lewis 1969 *A selection of Greek historical inscriptions to the end of the fifth century B.C.*
Mikalson, Jon D. 'Religion in the Attic Demes', *American Journal of Philology* 98 (1977) 421-35
Mitsos, Markellos 1976 'Ανάθημα δεκάτης αἰγῶν', *Proceedings of the 1st International Conference of Peloponnesian Studies* II: 20-22
Moretti, Luigi 1967 *Iscrizioni storiche ellenistiche* I
National Statistical Service of Greece 1966 'Αποτελέσματα τῆς ἀπογραφῆς γεωργίας-κτηνοτροφίας τῆς 19 Μαρτίου 1961, I
Osborne, Robin 1985 *Demos: the discovery of classical Attika*
Parker, Robert 1983 *Miasma. Pollution and purification in early Greek religion*
Payne, Sebastian 1973 'Kill-off Patterns in Sheep and Goats. The Mandibles from Aşvan Kale', *Anatolian Studies* 23: 281-303
1985 'Zoo-archaeology in Greece: a reader's guide' in Nancy C. Wilkie and William D.E. Coulson, edd. *Contributions to Aegean Archaeology. Studies in honor of William A. McDonald* 211-44
Pfister, Friedrich 1951 'Die Reisebilder des Herakleides', *Akademie der Wissenschaften zu Wien, Sitzungsberichte, Philosophisch-historische Klasse* 227, 2
Pritchett, W.K. 1953 'The Attic Stelai, Part I', *Hesperia* 22: 225-99
1956 'The Attic Stelai, Part II', *Hesperia* 25: 178-328
Puttkammer, Friedrich 1912 *Quo modo Graeci victimarum carnes distribuerint* (Dissertation Königsberg)
Rapp, George, Jr. and S.E. Aschenbrenner, 1978 *Excavations at Nichoria in Southwest Greece.* Volume I. *Site, environs and techniques*

Rappaport, Roy A. 1979 *Ecology, meaning and religion*
Reese, David S. 1984 'Faunal Remains from the Kommos Temples', (abstract) *American Journal of Archaeology* 88: 257
1985 'The Kition Astragali', Appendix VIII (C) in V. Karageorghis *Excavations at Kition* V, Part 2 382-91
forthcoming "Faunal remains from the Aphrodite attar, Athens" *Hesperia*
Rhodes, P.J. 1981 *A commentary on the Aristotelian Athenaion Politeia*
Richter, Will 1968 *Die Landwirtschaft in homerischen Zeitalter* (*Archaeologia Homerica* 2,H)
1972 *RE* s.v. 'Ziege', 398-433
Robert, Louis 1945 *Le sanctuaire de Sinuri près de Mylasa* I *Les inscriptions grecques*
Roesch, Paul 1985 'Les taureaux de bronze du Kabirion de Thèbes et l'écriture archaique béotienne', *Proceedings of the Third International Conference on Boiotian Antiquities* 135-51
Shear, T. Leslie 1984 'The Athenian Agora: Excavations of 1980-1982', *Hesperia* 53: 1-57
Sivignon, Michel 1976 'Frontier between Two Cultural Areas', in Dimen and Friedl: 43-58
Slicher van Bath, B.H. 1963 *The agrarian history of Western Europe A.D. 500-1850*
Sloan, Robert E. and Mary Ann Duncan 1978 'Zooarchaeology of Nichoria', in Rapp and Aschenbrenner, 1978: 60-77
Sokolowski, Franciszek 1955 *Lois sacrées de l'Asie Mineure* (L'École Française d'Athènes, in the series *Travaux et mémoires des anciens membres étrangers de l'école et de divers savants* Fasc. IX) = *LSAM*
1962 *Lois sacrées des cités grecques, Supplément* (Fasc. XI) = *LSCG, Suppl.*
1969 *Lois sacrées des cités grecques* (no fascicle number) = *LSCG*
Stengel, Paul 1879 *Quaestiones Sacrificales*
1910 *Opferbräuche der Griechen*
1920 *Die Griechischen Kultusaltertümer* ed. 3
Stroud, Ronald 1965 'The Sanctuary of Demeter and Kore on Acrocorinth Preliminary Report I: 1961-1962', *Hesperia* 34: 1-24
Topping, Peter 1976 'Premodern Peloponnesus: the land and the people under Venetian rule (1685-1715)', in Dimen and Friedl, 1976: 92-108.
Traill, John S. 1975 *The political organization of Attica: a study of the demes, trittyes and phylai, and their representation in the Athenian council* (*Hesperia*, Supplement XIV)
1986 *Demos and Tritty. Epigraphical and topographical studies in the organization of Attica*
Ventris, Michael and John Chadwick 1973 2nd ed. *Documents in Mycenaean Greek*
White, K.D. 1970 *Roman Farming*
Whitehead, David 1986a *The Demes of Attica 508/7- ca. 250 B.C. A Political and Social Study*
1986b 'The "Greater Demarchy" of Erchia', *The Ancient World* 14: 57-64
Wolff, Petra 1978 'Die Tierreste', in Bammer et al.: 107-15
Ziehen, Ludwig 1896 'Die panathenaïschen und eleusinischen ἱεροποιοί', *Rheinisches Museum* 51: 211-25
1899 'εὔστόν', *Athenische Mitteilungen* 24: 267-74
1906 J. von Prott and L. Ziehen *Leges Graecorum Sacrae* I *Leges Graeciae et Insularum*
1939 *RE* s.v. 'Opfer' 579-627

NOTES

1. Stengel, 1920: 105-6. The bulk of the evidence is consistent with this view but [Arist.] *Oec.* 2.20e, 1349b distinguishes σφάζοντες ἐπώλουν from ἱερόθυτα ἐποίουν.

2. On the Attic Dipolieia/Bouphonia and related rites and stories, see Durand, 1986. For a comparable prohibition in Rome, see Ampolo, 1980: 35, 46.

3. The fact that working cattle could be eaten at the end of their useful lives was one reason for their late replacement by equids, whose consumption was against the cultural code, on the Methana peninsula early in this century (Hamish Forbes, personal communication).

4. For the abbreviations *LSCG*, *LSCG, Suppl.* and *LSAM* see Sokolowski in the bibliography. For ML see Meiggs and Lewis.

The Erchia calendar was first published by Daux, 1963: 603-34. Cf. Dow, 1965: 180-210; *SEG* XXI.541, XXII.131. The calendar is entitled *Demarchia he mezon*, and it seems likely that a second, lesser list included rituals not performed every year and perhaps the notably absent rites for Demeter, cf. Mikalson, 1977: 421-35. For a review of the problem see Whitehead, 1986b: 57-64.

5. For a review of the evidence for 'Dog Burials in the Greek World,' see Day, 1984. It has been suggested on the basis of cut marks on the bones that dogs were consumed in the Middle Helladic I phase (ca 2100-1850 B.C.) of Nichoria in Messenia (Sloan and Duncan, 1978: 69). So too horses or asses, deer and small ruminants have been thought to have been sacrificed at the altar of Artemis at Ephesos (Wolff, 1978: 109-10 and Bammer, 1978: 150, 154). However the consistently wide variety of bones recovered in careful excavations shows that various processes have brought bones to the site, and that those occurring in small quantities are not an indication of normal practice on the site. See the warnings of Boessneck, 1973: 1, Payne, 1985: 225. Much food debris, not necessarily the result of sacrifice, is found in hearths and near altars. The deer that occur at a number of sites may have been killed in hunting and have had a portion or a limb dedicated as a first-fruit offering; it has not been demonstrated that in historical times live deer were normally killed in sacrifice (cf. Ziehen 1939: 591-92, *pace* Kadletz, 1976: 84-87; with Plut. *Ages.* 6, contrast the silence of Xen. *Hell.* 3.4.4 who is careful to refer to significant ritual). Fish and fowl were offered, the former primarily as cooked food, the latter as an inexpensive offering and as a specific offering for certain deities (see Reese, forthcoming). In a few, distinctive rites, such as those for Artemis Laphria at Patrai (Paus. 7.18.12-13), a variety of captured wild animals were burnt whole but we have no bones from these cult places and comparable actions have not been attributed to those sites that have yielded bones.

Outside of normal Greek sacrifice is the remarkable deposit of 234 units of 'nearly complete bodies or butchered segments' from 5 cattle, 33 sheep or goats (of which 8 were lambs or kids), 22 birds and 26 hen's eggs in the Maussolleion at Halikarnassos in Caria, Asia Minor from shortly before the middle of the fourth century B.C. (Højlund, 1981 and 1983.) There is no indication of burning, nor that any parts were reserved for eating. The relationship, if any, to known Greek sacrifices has yet to be established.

6. Dahl and Hjort, 1976; Koster and Koster, 1976; Koster, 1977; Payne, 1973 and 1985:228-9, who warns (personal communication) that particular circumstances and needs vary so greatly that 'mean age at death' is not a very useful concept; Sloan and Duncan, 1978: 65.

7. Daux, 1983: 150-74 and 1984: 145-52, but from autopsy I date the inscription to the 30s or 20s of the fifth cent., not the early fourth. On the demes and religion, in addition to Mikalson (1977), see Osborne, 1985, and Whitehead, 1986a.

8. A new study by David S. Reese (forthcoming) supersedes Foster, 1984, and Shear, 1984. In particular, earlier identification of the remains of doves, thought to be appropriate to Aphrodite, has now to be abandoned. The initial study also drew conclusions from astragali found in the deposit. A distinction needs to be made between astragali (the knuckle-bones of ruminants) and other bones found in sanctuaries. While it is probable that they originally came from sacrificed animals the fact

that they were used, whether worked or unmodified, for gaming and divination, and that they sometimes differ in species from the preponderance of bones at a site (e.g., the Artemision of Ephesos, Wolff, 1978), means that they cannot be assumed to belong to the same sacrificial assemblage as the bones of other domestic animals. For a brief review of the subject with bibliography, see Reese, 1985; the largest body of material comes from the Korykian Cave near Delphi, Amandry, 1984. Note that in addition to sanctuaries and graves, they were found in the Dark Age II settlement at Nichoria in Messenia where they constituted 20% of all bones, McDonald and Coulson, 1983: 26.

9. In the back of the front room, a treasury area, were found 71 astragali, several of which were burnt. Presumably they were offerings. At least 5 were of sheep and 3 of cattle. About half had been worked in some fashion. It is not necessary that the cattle from which three of the astragali derive were slaughtered at this sanctuary. In any case, the scarcity of cattle bones is not contradicted.

10. Detailed information on the southern Argolid in this period comes from the *catastico particolare* compiled for the Venetian authorities in the Peloponnese ca 1700 A.D., knowledge of whose contents I owe to the generosity of Peter Topping., cf. Topping 1976: 92-108.

11. Payne (personal communication) notes that cattle at the Kabeirion were probably twice or even four times as scarce as suggested by the raw count, which is affected by indiscriminate inclusion of ribs etc. Only at Altar U in temple B2-3 at Kommos are all other victims excluded (Reese, 1984), but the presence of *all* the bones of the victims suggests to me that it was not the site of normal sacrifice and that the rites performed were, therefore, probably specialized in function and limited in fequency.

12. A point I owe to Jens Skydsgaard; cf. Frazer, 1965: II, 432.

13. Despite the observation of the ability of cattle to tolerate a poorer quality diet than sheep or goats (Halstead, 1981: 323), historically there has been a preference for grazing them on meadows. Decasos, 1904: 124, noted that Elis, Acarnania and Thessaly, with tall summer grass, raise more cattle than most of Greece. Thessaly, Macedonia and Thrace (with buffalo) had the most in 1961, while they had almost vanished from the Eastern Peloponnesos (Kayser and Thompson, 1964: 322). In antiquity Arcadians exceptionally could get milk to drink from their cows (Plin. *N.H.* 25.94). It may be that meadows are necessary for profitable beef and milk cattle while lean working animals can be kept well enough in more arid environments, as the figures from the southern Argolid show. Although Decasos regarded beef production as unprofitable in Greece his figures for 1893 show 60,449 beef to 13,224 working cattle (126).

14. But Antiphanes II 84 Kock, ap. Athen. I, 43C, lists bread, figs, *boskēmata*, wool, myrtle berries, thyme, wheat and water as products of Attica. The juxtaposition with wool may suggest the animals are sheep.

15. Pritchett, 1953: 272, VI.68-73, and 1956: 255-60. Another property had a pig stye and a cow barn, 1953: 265, V.39-40.

16. This is the situation around Athens in this century. The density of cattle according to the 1961 census (National Statistical Service of Greece, 1966: I, Table 3.1) in the territory of ancient Attica (the area of the capital together with Attica outside of the capital but without the Megarid and non-

adjacent *eparkhies*) was 0.23 head per hectare, in the Megarid 0.42, while in Boiotia and southern Euboia combined it was only 0.12. This reflects the demand for cow's milk and beef in the urban area of the capital and is made possible by fodder brought in from outside the region.

17. Böckh, 1886: I, 94; Pritchett, 1956: 255; Richter, 1968: 41. Dikaiopolis restores conditions that existed before the war so that he can do business in the market with Boiotians, Peloponnesians and Megarians, one of whom is eager to sell his daughters as pigs (Ar. *Ach.* 719-22, 764-73). One could wish for more solid evidence. Curiously the traveller Herakleides in the Hellenistic period comments on the field, garden and orchard crops of Attica and Boiotia but makes no mention of animals in either region (Pfister, 1951).

18. At Nichoria pigs contributed the most protein (30%) after cattle and were the largest number of individuals after sheep. Their proportions were greatest, 35%, in the populous periods of the Late Helladic (through LHIIIA2) and then dropped to 23%. At all times 95% were killed before they were 3 years old, 53% between 1 and 2 years, and 25% at less than 1. Most of the smallest animals were cooked whole (Sloan and Duncan, 1978: 67-8). Ritual may have required only their blood. Their peak numbers in the Late Helladic suggests intensive rearing and/or the availability of hands to watch over them in the countryside. When space was available for extensive herding in the Dark Age more attention appears to have been given to cattle.

19. Loukopoulos, 1930. Campbell, 1964. Kavadias, 1965. Koster and Koster, 1976. Koster, 1977.

20. An incomplete calendar from Erythrai in Asia Minor, *LSAM* 26, second century B.C., has 18 *teleia* to 8 *galathena*, all sheep, to judge by the prices.

21. Most of the bones from the altar of Aphrodite in the Agora of Athens may also be from very young sheep/goats (Reese, forthcoming). Dr. Giraud Foster kindly informed me that very young sheep and goats also predominate at the sanctuary of Apollo Hylates at Kourion on Cyprus.

22. Payne, 1973: 302, notes the relatively high proportion of sheep and goats killed at Aşvan Kale in Anatolia between the ages of 2 and 6 during the Hellenistic/Roman periods and very cautiously suggests a concentration on the production of high-grade wool. The pattern resembles what we seem to have in classical Greece. Wool was certainly a concern for the Greeks, but perhaps Greek sacrificial practices had been adopted in Anatolia in these centuries.

23. The Attic calendar of the classical period was capable of considerable divergence from the agricultural cycle to which much of the ritual calendar was at one time pegged, but I am assuming, perhaps rashly, that overall it did not distort radically the earlier sequences. On the relationship of the agricultural calendar to agricultural cults, see Brumfield, 1981. For simplicity I equate Hekatombaion, the first month of the Attic year, with August, rather than July/August.

24. Perhaps Zeus Horios, who receives a piglet in January, oversees disputes that arise between demes over winter grazing in the lowlands.

25. Harold Koster (to whom I owe much good advice, imperfectly reproduced in this paper) informs me that in the present day the early summer is an excellent time to cull even pregnant ewes since a late pregnancy with poorer grazing in prospect increases the likelihood of the ewe's falling sick.

26. A detailed note in Rhodes, 1981: 650. A variant (Schol. Arist. *Eq.* 660) that says the vow was originally for cows but changed to goats is unlikely to be correct since female goats were especially Artemis's animal.

27. Tegea: *LSCG* 67 = *IG* V.2,3; Georgoudi, 1974: 178-80. Cf. *LSCG* 105 = *IG* XII.5,2 (Ios). Porph. *De Abstinentia* 1.25, on Kyzicos. Flocks of Helios, in Lakonia, *Hom. Hymn Ap.* (III) 411-13 (cf. *Od.* 12.129-30). Herodotus (9.93) tells of sacred sheep being herded at Apollonia in the northwest, a service performed by the leading men. Livy 24.3.3-7, Hera at Kroton.

28. I am assuming all adults were full grown, lambs and kids were half the weight of adults and piglets weighed 3 kg., all but the last probably being too generous, and I make no distinction between males and females. 20 sheep at 20 kg. = 400 kg., 4 lambs at 10 kg. = 40 kg., 11 goats at 20 kg. = 220 kg., 1 kid at 15 kg. = 15 kg., 5 piglets at 3 kg. = 15 kg., for a total of 815 kg. less 1 lamb (10 kg.) and 3 piglets (9 kg.) in holocaust = 19 kg.

29. Hansen, 1986, argues for a figure of ca 30,000 Athenian male citizens actually living in Attica in the second half of the fourth century B.C. With a hearth multiplier of 5 this raises the estimate of the total citizen population by one half and lowers the amount of meat available to each person from sacrifices by the same fraction. It is, of course, unrealistic to limit the recipients of meat to the citizens. Slaves might get much less but [Aristotle] (*Oec.* 1.5.6, 1344B) speaks of sacrifices and recreation being intended more for them than for the free, perhaps in the sense that the free had more opportunities. Metics participated formally in some festivals in Athens (*IG* I^3 82, 23 = I^2 84), were members of their own organizations (Ferguson, 1944), held sacrifices at home (Plato *Rep.* I, 328C) and could buy meat. But they were largely in or close to town and had access to the different pattern that prevailed there. It is also likely that fewer than the total number of deme members actually benefited from the local sacrifices since many by the fourth century would have lived in Athens. Erchia's quota on the Council was 6-7 members, according to Traill, 1986, 127, n. 15.

30. I should note that Payne regards the total sacrifices I have estimated as remarkably low in relation to the likely surplus of animals available to communities of this size. I agree, and no doubt many more animals were killed and eaten than the sacrificial regulations suggest. But we would need to postulate vastly greater consumption of meat from private sacrifices than we are aware of, or routine butchering without sacrifice, for which there is no solid evidence, in order to approach modern levels of meat consumption.

31. Pritchett 1956: 257-8, doubts the costliness of beef and, as cattle provided the most meat and was sacrificed in great numbers in Athens, that may have been the case when it could be purchased from butchers in Athens instead of being bought on the hoof and sacrificed. However, we have no way of knowing how frequently private persons would have done so. A comic complaint about a cheap 10 drachmai sheep (below most fourth century prices for sheep, Pritchett, 1956: 259-60) is made to point a contrast with the other accoutrements of a feast (Menander fr. 264 Koerte; whatever the value of the victim, the gods of comedy made much of men's Promethean selfishness in the division of the animal). Robert, 1945: 48-9, notes that when a specific weight of meat is assigned in inscriptions (most examples are Hellenistic in date) it is not less than one mina (0.431 kg. on the Attic-Euboian standard) and often more, when serving as honours and perquisites of office. We do not know how many such portions any one person might receive in a year. But a single portion of a mina's weight approaches what we estimated a deme member of Erchia might get from the sacrifices of the Greater Demarchia in a year's time.

32. Moretti, 1967, N. 43 + *SEG* XX.405, XXXI.328. Cf. *IG* IV.[2] 1.70, 71 and 72. A man who may well have been the ancestor of one of the Epidaurian guides for the arbitration commission for the dispute with Hermion dedicated a small altar as 'a tithe for the goats' to the Anakes (Dioskouroi) at Epidauros town (Mitsos 1976: 20-22).

ROME ARCHAÏQUE: UNE SOCIÉTÉ PASTORALE?

Les Romains de l'époque historique se faisaient une idée assez simple de leur origine: les fondateurs et les premiers habitants de la ville avaient étés des bergers. D'ailleurs le *dies natalis* de Rome coïncidait, comme tout le monde le sait, avec le *Parilia* ou *Palilia* du 21 avril, une fête de bergers. Parmi les textes il suffit de citer un passage bien connu de Varron (*RR* 2.1.9):

> Romanorum vero populum a pastoribus esse ortum quis non dicit? Quis Faustulum nescit pastorem fuisse nutricium qui Romulum et Remum educavit? Non ipsos quoque fuisse pastores obtinebit quod Parilibus potissimum condidere urbem? etc.

Et Varron est seulement un parmi les nombreux auteurs qui parlent des fondateurs de la ville comme des bergers.[1] Il insiste sur ce point et donne aussi des étymologies pastorales des noms de lieux de la Rome archaïque et d'autres mots.[2] Varron avait adopté la théorie grecque des genres de vies (*bioi*), développée notamment par Dicéarque; surtout il voulait privilégier et encourager le développement de l'agriculture et il se proposait de montrer comment le pastoralisme et l'élevage du batail étaient historiquement liés à une phase culturelle arriérée, depassée.[3]

Mais les anciens croient volontiers qu'après la fondation de la ville, on dirait même avec la fondation, on va créer une société fondée sur l'agriculture (voir par exemple Denys d'Halicarnasse 2. 28). Pline l'Ancien, dans un texte important dedié à l'économie agricole de Rome archaïque (*Nat. hist.* 18. 2. 6 - 3.14) s'est efforcé de montrer la place de l'agriculture dans les premiers temps de la ville. Lui aussi, évidemment, chante les louanges de l'agriculture italienne, mais il se propose de montrer que la vie rurale était exercée et éstimée chez les prémiers Romains (*Nat. hist.* 18. 1.5: *honosque apud priscos maximus fuerit*); et précisément à cause de ça il y avait dans cette époque reculée abondance de récoltes, rendements élevés du blé et bas pri des céréales (voir 18. 4. 15-21). Ces pages, comme celles de Varron, ont évidemment un caractère rhétorique et antiquaire, mais elles contiennent aussi des données précieuses; il faut souligner un aspect significatif: les références au bétail sont toutes faites *en fonction de l' agriculture*, alors que chez Varron il y avait un raisonnement assez contradictoire (c'est-à-dire les deux activités sont soit associées soit séparées; cfr. *RR* 1.2.15 et 2.1).

Ce n'est pas le lieu pour traiter de la vision pastorale des origines de Rome chez les anciens; j'y reviendrai à la fin de cet exposé parce que cette image, cette Arcadie romaine – qui est d'ailleur très belliqueuse – a été utilisée par des savants pour confirmer la conception pastorale des origines. Quelques remarques sur l'histoire de la question de l'économie romaine archaïque peuvent nous enseigner encore beaucoup de choses.

Niebuhr, comme on le sait, avait interpreté toute l'histoire romaine archaïque à la lumière de la question agraire. À partir de son texte sur les lois agraires – resté longtemps inédit – jusqu'à l'*Histoire romaine*, les bergers ne jouent aucun rôle. Mommsen, dans des très belles pages de son *Histoire romaine* avait écrit que depuis longtemps les peuples italiques avaient dépassée le stade de l'économie pastorale et que l'agriculture restait le pilier de toutes les sociétés italiques, sabellique, étrusques et latines ('Der Feldbau blieb der Grundpfeiler aller italischen Gemeinden, der sabellischen und der etruskischen nicht minder als der latinischen'). Pour Mommsen il n'existaient plus en Italie des véritables tribus de bergers, mais partout on exercait ensemble l'agriculture et l'élevage, de manière et dans une mesure très différentes.[4] Il soutenait que même à l'époque des origines de Rome il y avait eu liaison étroite, interdépendance, des deux activités; et cela même dans un régime de communauté des terres. Une conception analogue se trouve dans la *Römische Agrargeschichte* de Max Weber (publiée à Stuttgart en 1891) dans le troisième chapitre à propos de l'*ager publicus* et de l'*ager compascuus*.

Une appréciation très différente du rôle de l'élevage revient de temps en temps chez les savants qui s'opposaient aux idées de Niebuhr et de Mommsen; je renvoi par exemple aux pages de Bachofen dans la *Geschichte der Römer* de Bachofen et de Fr.D. Gerlach, où l'on trouve les données sur l'importance – statistique et économique – de l'élevage dans le Latium du XIXème siècle.[5]

La thèse d'un pastoralisme romain originaire gagne du terrain dans les milieux historiques au début du XXème siècle. Ettore Pais, ancien élève de Mommsen, affirmait le 'carattere originariamente pastorale, anziché agricolo, dei più antichi Romani'; mais il ajoutait que pendant la deuxième partie de l'époque royale 'se manifestent les éléments constitutifs de la société romaine, agricole et militaire en même temps'.[6]

Mais il faut attendre l'*Essai sur les origines de Rome* de André Piganiol (Paris 1917) pour trouver une exposition détaillée de la thèse du caractère pastorale des origines de Rome, même s'il y a des précédents (qui à vrai dire n'ont pas eu beaucoup d'importance).

Dans son livre, très riche d'hypothèses fragiles mais aussi d'une étonnante connaissance des études dans le domaine d'autres sciences humaines (la sociologie, au sens large qu'avait ce terme à l'époque, l'archéologie), Piganiol éxpliquait la naissance de Rome comme le produit de la fusion de deux sociétés différentes, auxquelles correspondent deux cultures, deux économies et deux races opposées et, en particulier, deux modèles de habitat. Selon Piganiol les bergers indoeuropéens s'étaient fixés dans le *Latium vetus* au début de l'age du fer et y avaient trouvé des agriculteurs méditerranéens. Ce schéma dualiste était enrichi par la théorie des modèles d'habitat, une typologie des peuples de pâtres et d'agriculteurs établie par Meitzen.[7] Ce savant avait cru pouvoir déterminer trois types d'habitat, coïncidants avec trois populations et systèmes de propriété de la terre: a) fermes dispersées (Celtes; propriété individuelle); b) villages avec redistribution périodique des terres (Germains; coexistence de propriété privé et d'exploitation collective des terres publiques); c) villages circulaires fortifiés (Slaves; propriété collective et économie pastorale).

Piganiol appliqua ce schéma à Rome archaïque; le premier modèle était utilisé pour les Sabins et les peuples sabelliques; le troisième pour la première cité du Palatin; un autre, un compromis entre la propriété individuelle et le comunisme, se serait developpé sous l'influence réciproque et grâce à la fusion des deux types précédents, donnant origine au régime agraire romain avec les *heredia* (*bina jugera*) et les terres publiques communes.[8]

L'opposition aux thèses de Mommsen est très nette. La position de Piganiol n'a aujourd'hui qu'un intérêt historiographique, mais à l'époque ce livre représentait un grand effort pour insérer explicitement les problèmes des origines de Rome dans les débats sociologiques, qui en même temps comprenaient les grandes discussions du XXème siècle sur l'histoire de la propriété de la terre. On y retrouve aussi l'écho des recherches sur la typologie des habitats classiques (comme celles de Kornemann). Du point de vue strictement philologique, et même si l'on réfléchit aux résultats de la recherche archéologique et historique successive, le livre de Piganiol était un échec. On est frappé par l'usage très libre des sources littéraires et par la recherche obstinée d'éléments ethniques et de dualismes, de couples d'opposition.[9]

La mise en valeur du pastoralisme des origines a été accepté par plusieurs savants dans notre siècle, mais dans le cadre de théories générales très différentes. Il est impossible de les mentionner ici. Je me limite à signaler qu'on retrouve souvent ce rôle dominant de l'élevage parmi les oeuvres des juristes, soit parmi ceux qui considèrent l'élevage du bétail comme la base économique nécessaire de la *gens* romaine, vue comme organisme politique qui precède la cité (F. De Martino), soit parmi ceux qui expliquent les luttes entre patriciens et plébéiens comme l'opposition entre les propriétaires de bétail et les agriculteurs qui pratiquent une culture intensive du sol (A. Guarino; cfr. aussi L. R. Ménager).[10] De toute façon, la thèse d'une communauté de bergers, ou fondée sur l'élevage, est très répandue aujourd'hui.[11]

Et pourtant si l'on avait examiné plus attentivement les données des fouilles relatives à l'utilisation des céréales et des végétaux et à la faune de Rome primitive, le rapport réel entre élevage et agriculture aurait été jugé de façon différente. Il faut dire que les travaux de E. Gjerstad ont traité ce problème rapidement, mais sans tomber dans des vues unilatérales ou préconçues, parce que son analyse était fondée sur les données provenantes des fouilles.[12] Moi-même j'ai essaié de donner un tableau d'ensemble en utilisant différents types de sources pour la période X-VIIème siècle av. J.-C., dans le cadre d'une enquête sur le milieu et sur le paysage agraire du Latium.[13]

D'autres travaux récents qui revêtent une grande importance ont mis en lumière la composante pastorale des sociétés de l'Italie ancienne, soit pour des périodes préhistoriques[14] soit pour l'époque républicaine et impériale.[15] Ces recherches rendent nécessaire de réexaminer la documentation dont nous disposons pour Rome jusqu'au Vème siècle, dans le but de déterminer – si cela est possible – les rapports et les caractères de l'agriculture et de l'élevage archaïque.

Nous pouvons utiliser avec une certaine confiance quatre séries de données; il s'agit des données paléobotaniques, des ossements d'animaux, du calendrier romain le plus ancien et de la loi des XII Tables. Les deux dernières sources sont relativement récentes (VIème et Vème siècle av. J.-C.), mais elles ont une grande valeur documentaire et historique; les deux premières séries de données sont dans le cas de Rome assez nombreuses pour autoriser à les considérer comme une base statistique; d'autres recherches à Rome et dans le Latium sont à l'heure actuelle encore inachevées ou inédites et pourraient modifier les conclusions que je présente.

Les données relatives à la faune

Les trouvailles proviennent des vieilles fouilles de la nécropole du Forum romain (S-VIIème av. J.-C.) et de lieux de culte, c'est-à-dire du dépôt votif du *lapis niger* dans le *comitium* (dans un contexte archéologique que appartient en grande partie au VIème siècle av. J.-C.) et surtout de l' *area sacra di S. Omobono* au *Forum boarium* laquelle comprenait un lieu de culte du VIème siècle (où, pendant le même siècle, un temple fut bâti et refait), détruit vers 500 av. J.-C. et remplacé au IVème par deux temples jumeaux.

Si l'on considère seulement le bétail d'élevage, nous disposons de 16 contextes funéraires et de deux en provenance de sanctuaires; jusqu'à present on ne dispose pas pour Rome de données relatives à des cabanes ou des maisons. Cela signifie que notre documentation est le produit de repas et d'offrandes funèbres et de sacrifices.

Or dans les tombes il y a une nette majorité de porcs (12 attestations), beaucoup moins d'ovinés (trois présences seulement) et de boeuf (une fois). Dans les lieux de cultes on constate une présence de bovins. A S. Omobono en particulier les sacrifices comprenaient les trois animaux des *suovetaurilia*, mais il s'agissait des vaches, truies et brebis gravides (selon un rite bien connu par le rituel des *Fordicidia*, qui prescrit le sacrifice d'une *bos forda*, une vache pleine).

Une base statistique satisfaisante, la présence d'associations stratigraphiques et la qualité de l'étude des découvertes pourraient permettre d'avoir un cadre complet de l'utilisation du bétail dans les différentes périodes. Mais les femelles pleines sont attestées toujours dans les couches associées au sanctuaire; en général il parait que les bovins sont plus nombreux dans les périodes plus anciennes et que les porcs deviennent de plus en plus nombreux après l'époque archaïque.[16]

Sous réserve de la particularité des contextes archéologiques, on peut donc affirmer que les Romains étaient d'un grand conservatisme pour ce qui concerne le bétail élevé pour l'alimentation. Ici les porcs ont la première place, comme à l'époque républicaine et impériale, suivis par les brebis et les moutons. Le boeuf, rare dons les repas funèbres, est plus fréquent dans les sacrifices (il s'agit du *bos brachyceros*). Ce cadre correspond bien à ce que l'on connait de l'alimentation romaine et des buts de l'élevage; il correspond aussi dans ses grandes lignes (et *grosso modo*) aux données sur la faune de l'Italie centrale à partir de l'âge du bronze et du fer.[17]

A ce propos je souligne les points les plus importants.

a) Le porc prévaut soit à Rome (dans les tombes datant d'entre le Xème et le VIIème siècle, et en général dans l'époque historique) soit dans un centre comme Narce (situé dans la région habitée par le peuple de Falisques, très voisin des Latins). Ici le tableau de l'alimentation semble rester invarié pendant plusieurs siècles. Près de Vulci, dans le lieu dit Sorgenti della Nova (dans la vallée du fleuve Fiora) à partir de l'âge du bronze final jusqu'au VIIIème siècle av. J.-C., et à Cures (la ville des Sabins, étroitement liée à Rome archaïque selon les traditions romaines) entre le IXème et le VIIème siècle le tableau est analogue. Il faut dire que dans ce dernier centre le porc et les ovines (ou les chèvres) sont en majorité numérique, mais le boeuf prévaut si l'on considère la quantité de viande qu'il pouvait donner.[18]

b) Il y a des variations possibles entre un centre et l'autre même dans les limites du *Latium vetus*. Ainsi par exemple dans la nécropole de l'âge du fer de la Osteria dell'Osa (près de la ville latine de Gabii) dans les repas et les offrandes funèbres on trouve surtout des restes de moutons ou de chèvres.[19] La base statistique est encore mince et les rituels funéraires peuvent être différents qu'à Rome; mais il est bien possible qu'ici (plus près de Praeneste et de l'intérieur de la région) on élevait plutôt de petit bétail.

c) Le grand bétail était peu utilisé dans l'alimentation courante, mais il était présent dans les fêtes et les sacrifices. Les données des tombes sont en accord avec la prohibition très ancienne de manger la viande du *bos arator*.[20]

d) Les ovines occupent une place notable après le porc, mais l'ensemble des données me parait montrer qu' en général l'élevage n'était pas specialisé; on a écrit à ce propos qu'il revèle plutôt un compromis entre l'utilisation de tous les produits (comme la laine, les peaux, le lait etc.) et la viande qu'on pouvait en obtenir.[21]

Les données paléobotaniques

Les 21 contextes archéologiques antérieurs à la fin du VIIème siècle (17 tombes, un puits et 3 couches, tous dans le Forum romain) et nombreuses couches de différentes époques fouillées dans l'*area sacra di S. Omobono* ont révelé qu'il y avait, toujours associés entre eux, des céréales inférieures et des légumes. Les céréales sont le *triticum dicoccum* et (ou, selon le cas) le *triticum spelta* (le premier est le *far* des Romains), le *triticum monococcum* et *l'hordeum vulgare* (dans le proportions suivantes: 58%, 10% et 32%); les légumes sont représentés par la *vicia faba* (la fève) et le *pisum* (le pois).

Le fait le plus important à mon avis est qu'*ils sont toujours associés*, dans pratiquement tous les contextes archéologiques, en particulier dans certains tombeaux (C et II de la nécropole du Forum romain). Les graines étaient grillées selon une procédure utilisée encore par les vestales romaines en pleine époque historique. Le mélange des céréales (et des légumes) montre certainement qu'on les utilisait tous ensemble, et qu'on mélangeait aussi les farines. La fréquence des associations est telle qu'on doit penser aussi que les premiers Romains mélangeaient également les différentes éspèces dans la cultivation, dans les semailles. Ce procedé est attesté encore à l'époque classique et il a été étudié plus particulièrement dans l'agriculture du moyen âge par Montanari, grâce aux nombreux contrats agraires de l'Italie du Nord. Si cette thèse, que j'avais soutenue en 1977, et qui me paraît être confirmée dans différents sites de l'Italie centrale, est vraie, on peut tirer de là deux conclusions:

– l'agriculture avait encore un caractère de agriculture de subsistance, c'est-à-dire que les hommes voulaient avant tout obtenir un produit certain et constant – dans un certain degré – dans les conditions de temps les plus variées (sécheresse, humidité etc.);

– le prix d'un tel choix (ou d'une telle habitude) devait être, comme dans le moyen âge, un rendement très bas, comme on peut d'ailleurs le vérifier dans le Latium jusqu'au XIXème siècle.[22]

Le développement de l'arboriculture, avec la vigne et l'olivier, est plus récent. Le produit est attesté dans le *Latinum vetus* par des données paléobotaniques et par des matériaux archéologiques (vases destinés à contenir le vin, l'huile et l'huile parfumée) à la fin du VIIIème siècle et le début du VIIème siècle; l'apparition de la vigne et de l'olivier dans certaines régions de l'Italie semble être plus ancienne.[23] J'en parle brièvement parce que cela indique un retard dans le développement de l'agriculture romaine archaïque, qui semble s'enrichir à partir du VIIème siècle. Pour les céréales même il faut dire que le niveau et la qualité des espèces n'étaient pas élevés: elles sont comparables à celles qu'on connaît en Etrurie à l'âge du bronze, où pourtant on trouvait des céréales à caryopse nu (comme le *triticum aestivum* et le *triticum compactum*), qui paraissent absents à Rome. Mais l'association de graines différentes est commune même en Etrurie.[24]

Le témoignage du calendrier archaïque

Une évaluation approximative du rôle et du pois qu'avaient dans la vie sociale romaine l'agriculture et l'élevage peut être obtenue grâce au calendrier romain plus ancien, celui qui est indiqué sur le *feriale* des époques postérieures avec des lettres plus grandes, comme Mommsen l'a montré. Ce calendrier archaïque doit remonter au VIème siècle; de toute façon, il ne peut en aucun cas être plus récent que la moitié du Vème siècle. Et, comme on le sait, bien des fêtes qui y sont enregistrées appartiennent à une periode très ancienne. Or sur un total de presque 50 fêtes (45 jours indiqués sur les inscriptions; on peut augmenter de nombre jusqu'à 49 si l'on considère les fêtes qui coïncidaient avec les jours fixes commes les ides, ou comme l'*equus october* et la fête d'Anna Perenna) on en trouve seulement deux qui ont un caractère pastoral, les *Parilia* du 21 avril et les *Lupercalia* du 15 février. Beaucoup plus nombreuses sont les fêtes qui ont un caractère agraire: *Fordicidia* (15 avril), *Cerealia* (19 avril), *Vinalia* (23 avril et 19 août), *Robigalia* (25 avril), *Opiconsivia* (25 août), *Meditrinalia* (11 octobre); d'autres aussi présentent des aspects agraires, comme les *Liberalia*, les *Consualia* et les *Saturnalia* – au moins à l'origine – mais il y a des discussions entre les savants sur leur caractère. Il faut aussi tenir compte des fêtes qui ont rapport aux eaux et aux bois et, plus en général, au bon usage de la nature.[25]

Il est évident que numériquement le poids des fêtes pastorales est mince; elles ont une importance particulière dans le domaine civique, dans l'histoire de la religion romaine et dans les récits sur les origines. Mais dans le cadre de la vie romaine (de la 'vie quotidienne' pour ainsi dire), dans le rythme naturel, dans l'usage religieux de la nature, leur importance est réduite, comme le montre leur nombre.

Une liste complète des animaux sacrifiés dans les fêtes du calendrier archaïque serait sans doute très intéressante, mais malheuresement ces sacrifices sont souvent attestés par des sources assez tardives (comme Varron, Ovide et d'autres antiquaires); nous ne sommes pas certains qu'il n'y a pas eu un changement dans certains cas. Mais on peut affirmer que dans les rites, c'étaient surtout des ovines et des porcs qui étaient les victimes préferées. Je me limite ici à signaler les sacrifices qu'on célébrait régulièrement aux calendes et aux ides: le premier jour du mois la femme du rois des sacrifice (*regina sacrorum*) immolait à Junon *porcam vel agnam in Regia* (Macrobe, *Sat.* 1.15.19) et un *flamen* immolait aux ides une brebis, qu'on appelait *ovis idulis* (Macrobe, *Sat.* 1.15.16; cfr. Paul Diacre 93 L).[26]

Le calendrier archaïque nous donne donc l'idée d'une société fondamentalement agraire, comme d'ailleurs toute la religion romaine archaïque, où la présence de deux fêtes pastorales montre que cet aspect était bien présent; mais qu'il revèle une couche antérieure ou qu'il vienne d'un secteur de la vie économique archaïque, cette présence pastorale n'est jamais dominante.

Le témoignage de la loi des XII Tables

C'est un autre document complexe qui peut nous aider à contrôler les rapports entre l'agriculture et le monde des bergers; les Lois des Douzes Tables présentent évidemment des mises à jour plus récentes du point de vue linguistique et dans les références aux mesures de la valeur, donc même dans des aspects économiques. Mais – comme on l'admet généralement aujourd'hui – elles préservent l'essentiel de la législation du Vème siècle et peuvent donc réveler l'aspect général, le reflet juridique de la situation économique du Vème siècle av. J.-C. De plus elles contiennent des prescriptions qui ont une plus grande antiquité (dans le cas de la table X).[27]

Sans entrer dans les détails, ce qui nous amenerait à des discussions essentiellement juridiques, on peut se limiter à des observations générales:

a) Le bétail est pris en considération seulement pour les dommages qu'il apporte aux propriétés d'autrui et aux produits des champs. Sont explicitement attestées l'*actio de pauperie* (VIII, 6-7) qui s'en prend au dommage provoqué par des quadrupèdes (Fet. 246 L) et l'*actio de pastu pecoris* (VIII, 6 et 7) qui concerne le bétail qu'on amenait au pâturage et qui détruisait la récolte d'un autre (VIII, 7 et 9). La peine prévue était la même que celle qui était appliquée au voleur, c'est-à-dire la pendaison et la consacration à Cérès.

b) Les lois montrent la préoccupation de protéger les récoltes soit du pacages soit des vols, comme des enchantements (VIII, 8) et des incendies criminels ou accidentels.

c) Les arbres sont protégés (VIII, 11) et les vignobles sont mentionnées deux fois (VI, 8 et 9).

d) Les dispositions relatives à la durée prévue pour l'*usucapio* (la prescription acquisitive romaine) prouve clairement qu'on emploiait couramment la rotation biennale des cultures avec un an de jachère (VI, 3 'usus auctoritas fundi biennium est, ceterarum rerum omnium annuus est usus').

Donc au Vème siècle la loi entendait protéger sévèremment les champs et les récoltes des menaces – si on peut dire ainsi – du bétail; elle se préoccupait aussi de sauvegarder les vignobles. L'accent est mis sur l'agriculture (céréales et vigne) mais le bétail est bien présent. L'impression qu'on tire de certaines dispositions de la VIII table est donc que ce bétail vivait surtout sur la propriété et pouvait endommager celle du voisin, mais il est également possible qu'il provenait des *agri compascuus et publicus*.

Conclusions

Les quatre séries de données qu'on a considerées ne permettent encore d'élaborer un modèle économique certain; toutefois elles montrent la coexistence de l'élevage et de l'agriculture; cette dernière montre un caractère 'de subsistence' et apparaît très simple avec l'utilisation de céréales inférieures et de légumes associés, et donc de rendements qui n'étaient pas élevés; seulement les cultures plus spécialisées de la vigne et de l'olivier (probablement dans une mesure encore réduite à l'époque archaïque) attestent un progrès à partir de 700 av. J.-C. et la présence de la triade classique des cultures méditerranéennes.

Cette association d'agriculture de subsistence et d'élevage du grand et du petit bétail nous permet – je crois – d'utiliser le modèle de Slicher van Bath qu'on peut à juste titre appliquer au Latium archaïque.[28] Il permet de comprendre cette interdépendance des deux activités, qui étaient nécessaires l'une à l'autre, parce que le bétail était indispensable pour la fumure du terrain pendant la période de la jachère et que le boeuf représentait la source d'énergie animale nécessaire pour labourer la terre. Le modèle peut être présenté dans cette forme très simple:

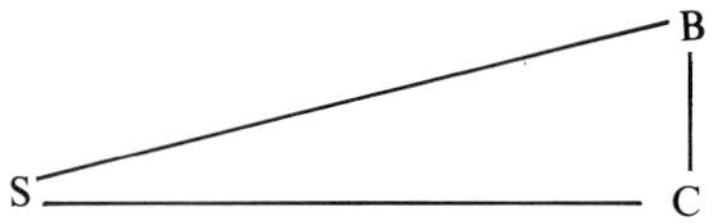

S = surface du terrain cultivé
B = bétail
C = fumier

Ce schéma pourrait être modifié si on ajoutait les éléments du milieu naturel (bois et forêts, marais, régime des eaux de surface) du Latium archaïque. Mais la structure fondamentale resterait la même.

Les caractères du milieu naturel et des sols du *Latium vetus* n'étaient pas très favorables à une agriculture avec des rendements plus élevés; cela peut expliquer le choix des céréales et leur mélange. Marais et bois, terrains humides au moins dans une partie de l'année, permettaient aisement l'élevage du porc; les pâturages des collines étaient favorables, au moins dans certaines saisons, au petit et au grand bétail. On ne peut rien affirmer de certain sur l'existence de la transhumance dans la période qu'on a considéré ici; mais pour l'époque préhistorique Barker et d'autres croient à des déplacement sur distances courtes ou moyennes. Des considérations topographiques, des noms de lieux comme celui du *Forum boarium*, un éventuel rapport de l'élevage avec le sel et les salines près de la côte et du Tibre rendent possible un déplacement à l'intérieur du Latium ou dans des territoires directement limitrophes.[29] Mais on ne dispose pas – et d'ailleurs cela serait presque impossible – de preuves de déplacement sur longues distances du bétail, comme ceux qui sont attestés dans d'autres périodes historiques.[30]

Si l'image pastorale des origines de Rome doit être ramenée à de plus justes proportions, s'il faut revenir, pour le Latium au moins, à la théorie mommsénienne en associant étroitement élevage et agriculture, comment expliquer le succès chez les anciens et chez les modernes de l'imagerie pastorale, de ces représentation des premiers Romains habillés en bergers?

Une réponse est évidemment à chercher dans les mythes et dans les légendes des origines de Rome, dans les deux fêtes dont on a parlé plus haut (les *Parilia* et les *Lupercalia*) et aussi dans la terminologie latine de la richesse.

Pour en rester à ce dernier point, les termes *pecunia, peculium, peculatus* etc. ont des étymologies bien connues et anciennes, dérivées de *pecus*.[31] Ce fait a été utilisé par les savants de l'antiquité et de notre époque comme une preuve certaine du pastoralisme romain. Mais E. Benveniste a mis en doute cette certitude, démontrant que dans les langues indoeuropéennes et tout particulier en latin c'était en fait le phénomène opposé qui se vérifiait. À partir de mots de signification générale on était passé aux termes spécifiques qui désignaient le bétail (à peu près comme le latin *capitale* a donné l'anglais *cattle* et le français *cheptel*).[32] Si la thèse de Benveniste est juste, il n'est plus possible soutenir que 'au début était le bétail', mais au contraire qu' à partir de désignations d'un caractère plus général les langues indoeuropéennes ont créé des mots spécialisés qui sont souvent devenus des termes pour désigner le bétail.

En ce qui concerne les légendes des origines on peut accepter peut-être ce qu'avait écrit Andreas Alföldi; selon lui le milieu du mythe des origines de Rome était constitué d'une société d'éleveurs de bétail, et il a souligné 'die zentrale Rolle der Hirtenwelt in den römischen Ursprungssagen'.[33] Et les parallèles qu'a rassemblés Binder sont aussi convaincants.[34] Mais cela vaut sur le terrain du mythe (et il ne faut pas oublier que le monde des bergers était considéré comme l'opposé du monde des cités, son renversement ou un stade précédent). Les deux autres conclusions (mais je dirais plutôt qu'il s'agit d'arguments *a priori*) sont absolument inacceptables et n'ont pas reçu de démonstration: un héritage eurasien dû à des peuples de pâtres et de guerriers, qui expliquerait ces mythes et la structure institutionnelle de la Rome archaïque, et l'arrivée des Latins dans le Latium à une époque qui n'était pas très éloigneé du début du premier millénaire: ils auraient été 'dediti all' allevamento del bestiame ed alla razzia come del tutto giustificate'.[35]

Cette deuxième affirmation a été démentie par la constatation de l'existence d'une continuité entre la culture de l'âge du bronze final (XIIème-Xème siècles av. J.-C.) et la culture du *Latium vetus* à l'âge du fer, une culture qui est certainement en liaison avec les Latins.

La documentation que j'ai examinée dans cette analyse, relative aux rapports entre agriculture et élevage du bétail à Rome, démontre que l'idée trop simple des *prisci Latini* éleveurs de bétail ne correspond pas à la réalité. Est-il possible qu'elle corresponde à l'économie du Latium au cours du deuxième millénaire? Je ne le sait pas et la réponse il faut la trouver dans les recherches des préhistoriens. Ce qui me paraît certain, c'est que cette image traditionnelle des premiers Romains, ce pastoralisme des origines, est à ranger parmi ces 'factoids' récemment évoqués par un historien de l'antiquité, ces choses qui rassemblent à des faits établis seulement parce qu'ils sont communément acceptées, mais qui n'existent pas dans la réalité.

UNIVERSITÉ DE PISE CARMINE AMPOLO

NOTES

1. Voir Liv. 1.4-6; 2.1-4 (*illa pastorum convenarum plebs*); 5.53.9 (*maiores nostri, convenae pastoresque*); Ovid., *Fast.* 4.809-10; Dion. Hal. 2.2.1 etc. Les autres références sont dans A. Alföldi, *Die Struktur des voretruskischen Römerstaates* (1974), 109-11 Faustulus est présenté par certains textes comme un porcher (*subulcus*, *suphorbos*); voir Dion. Hal. 1.79.9; Plut., *Rom.* 6 et *comp. Thes. et Rom.*; *origo gentis Romanae* 20.3. La coïncidence entre ce dernier texte, qui cite Fabius Pictor, et les deux précedents, indique que vraisemblablement déjà le premier historien romain faisait de Faustulus un porcher et non pas un berger.

2. Varro, *RR.* 5.53.54.164; voir aussi 5.36 et 37. L'étymologie de *Palatium* de *palare* est implicite dans Névius, fr. 27 Morel.

3. E. Gabba, M. Pasquinucci, *Strutture agrarie ed allevamento transumante nell'Italia romana (III-I sec. a.C.)* (1979) 43-4. La plus ancienne représentation de la vie pastorale opposée à l'agriculture et à la civilisation est l'image mythique de Polyphème et des Cyclopes dans l'*Odyssée* (9, 105-24; 216-23; par opposition voir 131-5). Les prémisses à la théorie des *Kulturstufen* de Dicéarque sont dans la *Politique* d'Aristote (1.1256a) et dans sa distinction des differents *bioi*; mais Dicéarque les transforme en stades. Voir en dernier lieu G. Bodei Giglioni, 'Dicearco e la riflessione sul passato', *RSI* 98 (1986), 629-52.

4. Pour les aspects agraires dans l'oeuvre de Niebuhr voir A. Momigliano, 'Alle origini dell'interesse su Roma arcaica: Niebuhr e l'India', *RSI* 92 (1980), 561 ss. = 7° Contributo (1984), 155 ss.; A. Heuss, *B.G. Niebuhrs wissenschaftliche Anfänge* (1981). La citation de Mommsen dans le texte est tirée de la *Römische Geschichte* ed. 6 (1874) 181; cfr. 20 avec la note 5. Pour la théorie agraire de M. il faut tenir compte du fait qu'elle est influencée par la thèse d'une période d'unité de Latins et Grecs bien avant l'époque historique, une période au cours de laquelle ils auraient déjà développé l'agriculture. Sur les problèmes agraires et les discussions sur les formes de la propriété P. Grossi, *Un altro modo di possedere* (1977) et L. Capogrossi Colognesi, *La terra in Roma arcaica* I (1981).

5. Les *Beiträge* de Bachofen au livre publié en 1851 sont réimprimés dans ses *Gesammelte Werke* I (1943); cfr. 79 ss. et 122: 'Wie zu den Zeiten der Gracchen, so bildet noch heute die Viehzucht Zweck und Grundlage der ganzen Kultur. Alle Verhältnisse des Landes, natürliche wie geschichtliche, führen auf diese hin.' Voir aussi les données de l'enquête de De Tournon de 1814, citées par Bachofen: dans le Latium il y avait alors 700.000 brebis et moutons avec 8.000 bergers et 100.000 bovins.

6. E. Pais, *Storia critica di Roma* 1,2, 2 ed. (1913) 645 = *Storia di Roma*, 2 (1926) 288.

7. A. Meitzen, *Siedelung und Agrarwesen der West- und Ostgermanen, der Kelten, Römer, Finnen und Slawen* I 1895. Sur les discussions du XIXème siècle autour des formes de propriété voir les oeuvres citées à la note 4.

8. Piganiol est resté fidèle à ses idées; voir *La conquête romaine*, ed. 2 (1962), chap. 5. Sur le débat entre 'morphologie sociale' et 'géographie humaine' dans les premiers decennies du siècle, voir L. Fevre, *La terre et l'évolution humaine* ed. 2 (1924) 39 ss. (référence à Meitzen à la page 55). Un exemple de l'usage de critères ethniques dans G. Oberziner, *Origine della plebe romana* (1901), en particulier 154-8.

9. Voir J.-Cl. Richard, *Les origines de la plèbe romaine* (1978) 61 ss.

10. F. De Martino, *Storia della costituzione romana* 1, (1958) 1, 26, 39-43.; *Storia economica di Roma antica* I (1979) 1-4 ('pastori che stanno scoprendo l'agricoltura'); A. Guarino, *La rivoluzione della plebe* (1975) 60-1 et *passim*; L.R. Ménager, 'Nature et mobile de l'opposition entre la plèbe et le patriciat', *RIDA*, 7 sér., 19 (1972) 367-98.

11. R. Besnier, 'L'état économique de Rome au temps des rois', *RHDFE*, 13 (1934) 405-63; L. Clerici, *Economia e finanza dei Romani* I (1943). Parmi de nombreux livres récents je signale la publication très générale de R.M. Ogilvie, *Early Rome and the Etruscans* (1976) et surtout les travaux de A. Alföldi (voir note 1 et plus bas dans le texte).

12. E. Gjerstad, *Legends and Facts of Early Roman History* (1967) 19; *Early Rome* 1-6, (1953-73) *passim*.

13. 'La formazione della Città nel Lazio', *DArch*, n.s. 2 (1980) 15ss.

14. G. Barker, *Prehistoric Territories and Economies in Central Italy* dans E.S. Higgs (ed.) *Paleoeconomy* (1975) 111 ss.; *Landscape and Society: Prehistoric Central Italy* (1981); *Prehistoric Farming in Europe* (1985) et beaucoup d'autres travaux qu'il est impossible de citer ici.

15. Gabba, Pasquinucci, *Strutture* (n.3).

16. G. Ioppolo, 'I reperti ossei animali nell'area sacra di S. Omobono', *RPAA* 44 (1971-72) 3 ss.

17. 'La formazione' (n.13).

18. G. Barker, M.H. Jarman dans T.W. Potter, *A Faliscan Town in South Etruria* (1976) 295 ss. et 308 ss.; *Sorgenti della Nova. Una comunità preistorica e il suo territorio nella Etruria Meridionale* (Catalogo della mostra, Milano 1981) (1981) 269 ss. (*sus scrofa* 60; boeuf 13; 'caprovines' 17 et un grand nombre de sangliers). Pour *Cures*: M. Ruffo dans *Archeologia Laziale*, 7, 1, (1985), 84.

19. *Ricerca su una comunità del Lazio protostorico* Catalogo della mostra (1979), 109.

20. 'La formazione' (n.13).

21. Pour *Cures* on a écrit que 'L'età di morte degli animali, orientata verso individui adulti, la piccola taglia ed il rapporto del numero di presenze tra tutte le specie (ovini, suini, e bovini) stanno ad indicare che il livello dell'allevamento era assai poco specializzato': Ruffo, n.18.

22. Les données dans 'La formazione' (n.13). Cfr. M. Montanari, *L'alimentazione contadina nell'Alto Medioevo* (1979) 109 ss. Pour l'opinion contraire voir De Martino dans *PP* 39 (1984) 241-62.

23. Pour le vin et l'introduction de la vigne voir maintenant M. Gras, *Trafics tyrrhéniens archaïques* (1986) 367 ss. avec la bibliographie. Pour l'olivier, il faut signaler l'opinion de F. Constantini dans *Archeologia laziale* 7, 1 (1985), 88 (avec n.18) qui croit à une origine beaucoup plus ancienne de la viticulture italienne; mais jusqu'à présent il n'y a pas dans le *Latium vetus* des témoignages qui confirment cette opinion. Pour la vigne dans l'Etrurie méridionale on signale maintenant la présence de plusieurs pépins de raisin dans le village du Gran Carro près de Bolsena (IX-VIIIème siècle av. J.-C.).

24. M. Follieri, dans *Sorgenti della Nova* (n 18), 263, signale qu'il s'agit fréquemment de petites fosses rituelles où l'on trouve ces céréales.

25. Voir les sources dans A. Degrassi, *Inscriptiones Italiae* XIII, 2. On peut trouver une liste de ces fêtes dans K. Latte, *Römische Religionsgeschichte* (1967) 433 ss. Pour ce 'bon usage de la nature' voir G. Dumézil, *Fêtes romaines d'été et d'automne* (1975); J. Scheid dans *AION (archeol)* 2 (1980) 41 ss. La bibliographie relative aux *Lupercalia* est dans Alföldi, *Struktur* 86-7, n.1; à propos des *Parilia* il faut remarquer que dans ce jour *oves lustrantur*, comme le démontrent le *menologium rusticum Colotianum* et le *menologium Vallense*. Pour Alföldi, (*Struktur* (n.1), 111) les *Consualia* auraient un caractère pastoral, mais il suffit ici de renvoyer à Dumézil, *Le religion romaine archaïque* (1966) 162.

26. Cfr. Latte, *Religionsgesch.* (n.25). 375 ss.; C. Krause, *De Romanorum hostiis* (1893); S. Eitrem, *Opferritus und Voropfer der Griechen und der Römer* (1915).

27. Voir en général F. Wiaecker, 'Die XII Tafeln in ihrem Jahrhundert' dans *Les origines de la république romaine* (Entretiens de la Fondation Hardt 13) (1967) 293-356 (sur l'économie 308); A. Watson, *Rome of the XII Tables. Persons and Property* (1975). Pour les aspects économiques voir aussi les remarques de Clerici, *Economia* (n.11), 84-5. Pour la table X je renvoi à mon étude dans *AION* (*archeol*), 6 (1984) 71-102.

28. Je renvoie à 'La formazione' (n.13).

29. Pour le sel voir A. Giovannini, 'Le sel et la fortune de Rome', *Athenéum*, n.s. 63 (1985) 373 ss. (mais cfr. déjà Clerici, *Economia* (n.11)). L'ancienneté de la via Salaria est un fait éloquent.

30. Voir maintenant les essais publiés dans 'Il Tevere' dans *Archeologia Laziale*, 7, 2 (1985). Pour le rôle des bacs sur le Tibre, G. Cozzo, *Il luogo primitivo di Roma* (1936) et L. Adams Holland, *Ianus and the Bridge* (1961).

31. Les textes sont rassemblés par R. Thomsen, *Early Roman Coinage* 1 (1957). Pour les origines de la monnaie à Rome voir maintenant M. Crawford, *Coinage & Money under the Roman Republic* (1985) 17 ss.

32. E. Benveniste, *Le vocabulaire des institutions indoeuropéennes* (1969) 47 ss. Voir aussi G. Melillo, *Economia e giurisprudenza a Roma* (1978).

33. *Struktur* (n.1), 107 ss.

34. G. Binder, *Die Aussetzung des Königskindes. Kyros und Romulus* (1964).

35. Alföldi, 'La struttura politica di Roma nei suoi primordi', *RAL*, s. 8, 27 (1972) 307-33 (la citation est prise de la page 306).

LA PASTORIZIA NELL' ETA' TARDO-IMPERIALE IN ITALIA

1. Nelle *Res rusticae* di Varrone (36 a.C.) ritroviamo un'eco significativa della polemica sullo sviluppo eccessivo della pastorizia ai danni dell' agricultura (2.*praef*.4-5). Anche se la polemica si riferisce principalmente all'allevamento sul fondo agricolo, tuttavia finisce, poi, per coinvolgere l'intiero problema del pascolo. Difatti Varrone si riffà alle teorie antropologiche di Dicearco, secondo le quali la pastorizia aveva seguito ad una fase nella quale la società umana viveva della raccolta dei frutti forniti spontaneamente dalla terra, ed aveva preceduto la fase dell'agricultura. La pastorizia in espansione poteva quindi essere intesa come un arretramento verso un fase inferiore di civilità (*Res rust*.1.2.16; 2.1.3-4). La stessa storia romana arcaica serviva da modello: la fase pastorale, tipica delle origini, aveva ceduto il passo all'agricultura: ora succedeva l'opposto. Queste riflessioni non erano soltanto l'esito di una schematica e astratta teoria sullo sviluppo dell'incivilimento umano; esse risentivano di una ben più viva discussione e dei formidabili contrasti politici e sociali dell'età graccana. Tiberio Sempronio Gracco, tribuno della plebe nel 133 a.C., aveva svolto nei suoi discorsi in appoggio alla proposta di legge agraria proprio argomentazioni di questo tipo e precisamente calate nella realtà contemporanea (App. *BC* 1.30; 35-36; 40; 46; Plut. *Gracchi* 8.4). La larghissima disponibilità di terre che, a seguito delle confische operate da Roma a danno dei *Socii Italici* che si erano ribellati durante la Guerra Annibalica, erano divenute *ager publicus Populi Romani*; l'altrettanto ampia disponibilità di mezzi finanziari in mano alle classi alte romane (e poi anche italiche) come conseguenza delle guerre di espansione del secondo secolo a.C. insieme con un deciso spopolamento di vaste aree del Centro-Sud della penisola italica avevano condotto in molti casi ad una utilizzazione del suolo differente da quella tradizionale. Tendenzialmente andò declinando la struttura, tradizionale in questo ambiente italico, della piccola proprietà contadina, complementata dall' utilizzazione delle terre comuni e con una produzione varia destinata all'autoconsumo: e si svilupparono, secondo le condizioni naturali del suolo, aziende piu vaste con produzioni specializzate destinate alla commercializzazione (la 'villa' catoniana), oppure gli allevamenti su vasta scala: anche questi richiedevano iniziali, ingenti spese di impianto, ma erano altamente redditizi, anche per la produzione della lana. Ed è in questo periodo (II sec. a.C.) che fenomeni già esistenti e legati alla situazione ambientale, come la transumanza stagionale del bestiame da pascoli montani a quelli di pianura e viceversa, vennero aquistando nuove dimensioni e si razionalizzarono, come appunto dimostra la descrizione

varroniana nel libro II delle *Res rusticae*. (Se Catone nel suo *De agricultura* sembra non conoscere la transumanza è perchè questo sistema è estraneo all'azienda – tipo la lui descritta e all'area geografica nella quale questa è collocata; ma noi sappiamo che Catone conosceva bene l'alta redditività dell'allevamento svolto su terreni, anche publici, e communque estranei al fondo).

Bisogna anche rammentare che le molte e varie forme dell'economia silvo-pastorale erano ancora predominanti, e si mantennero tali anche in seguito, in large zone montagnose dell'Italia Centro-Meridionale. Questa realtà, connessa a condizioni naturale, geografiche e ambientali, era difficilmente soggetta nell'età antica a cambiamenti dovuti a mutamenti politici. A questa realtà erano collegati modi tipici di sfruttamento del suolo e soprattutto l'utilizzazione collettiva del suolo; le condizioni ambientali non facilitavano il passaggio generalizzato a forme di appropriazione privata del suolo. Si tocca qui uno degli aspetti più importanti, e generalmente ammessi, in una prospettiva di continuità delle condizioni socio-economiche di molte zone del Centro-Sud dell'Italia: la presenza costante di terre fiscali, o comunque rimaste estranee al processo continuo di trasfomazione in proprietà privata; la loro varia utilizzazione, in diversi contesti storici e regioali, è stata determinante nel quadro generale dell'agricoltura e dell'economia italica.

Orbene, le profonde transformazioni nell'utilizzazione del suolo italico nel corso del secondo secolo a.C., considerate da un punto di vista generale, rappresentavano, o potevano rappresentare, un grande avanzamento economico; esse, però conducevano a gravi crisi sociali, in quanto, fra l'altro, favorivano ulteriori regressi della popolazione (sollecitata all'emigrazione) e un declino del ceto contadino libero, nel mentre i nuovi indirizzi dell'economia agraria si avvantaggiavano della ricca disponibilità di mano d'opera schiavile. L'intendimento di Tiberio Gracco era appunto quello di ricreare il ceto dei piccoli proprietari contadini, ricuperando parzialmente l'agro pubblico occupato più o meno legalmente dai ricchi: in questa prospettiva si sarà anche previsto, là dove era possibile, e conveniente, un ritorno all'agricoltura di terreni ora destinati al pascolo. (Il grande storico inglese Toynbee ritiene, con notevole esagerazione, che il governo romano dopo la Guerra Annibalica abbia confiscato ai Socii ribelli le migliori terre coltivabili e che queste siano state di regola trasformate in pascoli: egli è portato in generalizzare il fenomeno dell'allevamento e, specialmente, di quello transumante). Che la tematica del contrasto fra agricoltura e pastorizia fosse nell'età graccana profondamente attuale e operante anche sul piano politico, è dimostrato dalla famosa iscrizione di Polla, in Lucania, dove un anonimo magistrato romano vantava la propria priorità (forse con intenzioni anti graccane?) nell'aver riconvertito il pascolo in colture arative (Degrassi, *ILLRP* I^2 454).

Il fenomeno dell'allevamento 'industrializzato' andò crescendo nel corso del primo secolo a.C. e le *res pecuariae* rappresentarono una delle forme principali di investimento delle classi ricche romane, tanto in Italia quanto nelle province (Cic. *Cluent.* 198; *De orat.* 2.284; Varr. *Res rust.* 2. *praef.* 6; 2.1.2). La straordinaria ampiezza delle mandrie e greggi possedute in età augustea da un ricco liberto è ricordata da Plinio (*NH* 33.135).

2. Questa teoria di un regresso economico e quasi antropologico, legato all'espansione della pastorizia, era naturalmente dipendente da concezioni e idealità sociali e politiche che identificavano nel piccolo proprietario contadino la base della società romano-italica. E da questo punto di vista la polemica era certamente giustificata, oltre che comprensibile. E tuttavia le nuove forme di pastorizia 'industrializzata' (che sono poi quelle veramente in espansione) non erano affatto alternative all'agricoltura: si trattava, anzi, di una specializzazione che si combinava con modi più avanzati di utilizzazione del suolo, che, tuttavia, scardinavano la struttura della piccola proprietà.

La situazione era molto diversa nell'area celtica della Cisalpina. Nella descrizione che Polibio dà, a metà circa del secondo secolo a.C., delle strutture sociali ed economiche della regione, vi è insistenza sull'importanza della pastorizia (2.17.8 ss.), beninteso nel vasto quadro della ricchezza agraria della intiera zona. Per quelle popolazioni, che non avevano ancora superato lo stadio della non-sedentarietà, le greggi erano oggetto, con l'oro, di proprietà personale forse in un diverso contesto per quanto riguardava la terra. Vi è sottintesa la fondamentale differenza con il mondo greco-romano, la cui base di civilità è rappresentata dall'insediamento urbano, che le popolazioni celtiche non avevano, in quell'età, ancora raggiunto. L'allevamento dei suini (2.15.1-6) era favorito dalla grande estensione dei boschi in pianura, che in seguito e lentamente andarono cedendo spazio alle culture.[1] Nella valle del Po il processo di romanizzazione, favorendo l'urbanizzazione e la sedentarietà dell'insediamento umano, superò la fase pastorale celtica, anche se l'allevamento conservò sempre un'importanza economica alta. Alla fiera del bestiame dei Campi Macri presso Mutina, già attiva agli inizi del secondo secolo a.C., affluivano in età varroniana acquirenti anche dall'Italia Centrale (Varr. *Res rust.* 2. *praef.* 6; Strabo 5.1.11). Anche la fiera cittadina di Cremona (Tac. *Ann.* 3.30.1; 3.32.2) avrà interessato largamente il bestiame. L'area fra Parma e Mutina era famosa per gli ovini e per la produzione laniera. L'importanza degli allevamenti, anche equini, di Patavium era notissima, anche perchè alimentava una fiorente produzione di vestiario esportata anche a Roma (Strabo 5.1.4 e 7). Anche però nel caso della Transpadana va rammentato che permanevano aree ancora dominate dall'economia della palude, e tali da condizionare anche l'agricoltura.[2]

3. Per tornare all'Italia Centro-Meridionale, e al processo di razionalizzazione economica che indubbiamente si verificò nel corso del secondo secolo a.C., non si insisterà mai abbastanza su di un fattore di natura economico-politica, che favorì

grandemente il mutamento nell'utilizzazione del suolo negli indirizzi dell'agricoltura, già avviato dalle confische post-annibaliche. Dalla fine del terzo secolo a.C. Roma poteva disporre delle ricche produzioni granarie della Sicilia e della Sardegna, alle quali si aggiunsero nel secondo secolo la risorse della Spagna e dell'Africa Cartaginese. Queste produzioni erano in certa parte dovute a Roma come tributo pagato dalle province. Il caso della Sicilia è sopratutto indicativo. Il governo Romano perseguì, con maggiore o minore consapevolezza, nei riguardi della provincia una tipica politica di sfruttamento, in pratica imponendo la monocultura cerealicola, il cui esito era appunto la città capitale. La facilità dei trasporti marittimi, al confronto di quelli per via di terra, favorì questo smercio dei prodotti provinciali e finì, col tempo, con il rendere Roma dipendente dagli approvvigionamenti d'oltremare. La Gallia Cisalpina, della quale si è fatto cenno più sopra, rappresentò un mercato chiuso e a sè stante, con prezzi molto inferiori al resto d'Italia, appunto perchè la sua produzione rimase sostanzialmente destinata al consumo interno.

Orbene la possibilità di ricevere pressochè gratuitamente il frumento siciliano favorì la conversione, diciamo quasi 'industrializzata' dell'agricoltura italica, almeno in parte. Vino e olio erano facilmente esportabili a Roma e anche, almeno fino al primo secolo a.C., nelle province. L'economia italica divenne così, in certo senso e pur senza alcuna precisa volontà del potere, complementare a quelle provinciali e sopratutto siciliana.

Si tocca così un altro punto centrale. L'agricoltura italica si venne riorganizzando in funzione del nuovo mercato 'imperiale' di Roma (così come gli antichi rapporti di alleanza fra Roma e le città italiche dopo la Guerra Annibalica si vennero modificando in accordo con la nuova posizione imperiale di Roma). La nuova dimensione imperiale della politica romana ebbe riflessi sull'economia italica (e, come si è detto, anche provinciale) che a noi oggi appaiono inevitabili. Già nell'antichità si ebbe la consapevolezza che l'impatto del dominio romano su di una economia tradizionale era dirompente; si parlò volentieri di una decadenza morale perchè si era attenti alle conseguenze del nuovo stato di cose sulla vita pubblica e privata cittadina. Per noi il trasformarsi dei modi di vita, delle realtà sociali ed economiche, di idealità e comportamenti deve essere inteso come conseguenza di una situazione geopolitica completamente nuova.

È in questo contesto storico che si deve intendere lo sviluppo della pastorizia nel secondo-primo secolo a.C., specialmente di quello transumante, che rappresentava una forma di razionalizzazione economica (come già si è detto) e certamente una fonte di più sicura e immediata redditività.

4. E' necessario qui ripetere, per chiarezza, che la transumanza nell'Italia Centro-Meridionale, come risulta da una vasta serie di dati archeologici e topografici, era ben più antica del secondo secolo a.C. Inoltre la più avveduta indagine antropologica ha elaborato persuasivamente il concetto che questi spostamenti di greggi

inter-regionali, che connettevano economie arcaiche differenti e complementari, non presupponevano nè richiedevano legami politici fra le popolazioni interessate: come altre forme di scambio o rapporto economico, la transumanza è un sistema prepolitico. Con il secondo secolo a.C. il fenomeno si trasformò completamente rispetto al passato, assumendo dimensioni e caratteristiche nuove. Non più greggi di piccoli proprietari fra di loro consorziati si muovono sui tratturi (anche se di questo aspetto rimangono ancora in età imperiale dei residui); si tratta ora di masse di bestiame soprattutto ovino che si muovano stagionalmente, organizzate da personale libero e schiavile. In questa industria i grandi proprietari hanno investito ingenti capitali.

In questa nuova situazione lo stato romano dovette intervenire, sia per garantire la libertà di transito sulle *calles* pubbliche, sia per mettere in piedi un pur semplice apparato burocratico-fiscale per riscuotere i pagamenti dovuti per l'uso dei pascoli pubblici, sia infine per garantire l'ordine pubblico. Il tradizionale allevamento transumante, ora praticato su vasta scala, conobbe allora un cambiamento nel senso che venne ad assumere quel carattere ufficiale e 'forzoso' che sarà sua caratteristica per larga parte dell'età antica e poi in altre fasi della sua storia secolare.

5. In altri termini dobbiamo dare un' interpretazione globale dell'economica italico-romana a partire dal secondo secolo a.C. Con l'assurgere dello stato romano a potenza principale nel Mediterraneo, la stessa città capitale acquistò aspetti sociali, dimensioni urbane e funzioni economiche conseguenti. Di recente il cosiddetto modello del von Thünen e la sua teoria della 'location' sono stati applicati all'economia romano-italica del secondo secolo a.C. da P.W. de Neeve[3] per mostrare come appunto l'economia italica venne ad adattarsi alle nuove esigenze del mercato imperiale romano. Come si sa, secondo il von Thünen le produzioni agrarie si vengono a collocare entro aree formanti cerchi, o meglio ambiti, concentrici attorno al nucleo di un mercato. La produzione di ogni zona è condizionata dai costi e dalla possibilità dei trasporti di queste produzioni al mercato. L'allevamento si colloca di regola nella fascia più esterna. Questo modello (qui rapidamente riassunto) ha un alto grado di astrattezza e tuttavia la teoria applicata a Roma raggiunge molti aspetti di verità. Essa deve essere integrata tenendo conto dei fattori politici e sociali dell'Italia nell'età postannibalica e soprattutto delle condizioni ambientali, tanto naturali quanto storiche, delle aree nelle quali, in Italia Centro-Meridionale, predominava l'allevamento: condizioni che, come si è detto, erano un serio ostacolo al cambiamento (in quelle aree, per esempio, il processo di urbanizzazione promosso dello stato romano dopo la Guerra Italica nel primo secolo a.C. attecchì poco e finì anzi spesso per fallire). Per ribadire il peso delle condizioni ambientali sul permanere nel tempo di modi tipici di utilizzazione del suolo, è importante notare quanto scarse siano per l'età tardo- repubblicana le attestazioni di proprietà private nelle aree delle

regioni centro-meridionali (Apulia, Calabria, Lucania, Bruzio), mentre sono ampiamente ricordate le presenze di greggi: una ulteriore conferma, sembra, del prevalente carattere fiscale delle terre in quelle zone, di un mancato (perchè anche non possibile) processo di appropriazione di quelle terre, di una loro tipica utilizzazione, anche se inserita in un più avanzato contesto storico-economico.

Si è insistito sulla teoria del von Thünen (per quanto applicabile all'Italia Centro-Meridionale, almeno con grande cautela) perchè il modello interpretativo elaborato in base ad essa sembra importante per capire anche *fino a quando* a *perchè* il fenomeno dell'allevamento transumante abbia mantenuto l'alto grado di efficienza che aveva raggiunto fra secondo e primo secolo a.C. Si è abbastanza d'accordo nel ritenere che il massimo fiorire della transumanza si colloca fra il secondo secolo a.C. e il primo d.C., vale a dire fino a quando durò la centralità economico-politica dell'Italia e di Roma. Di nuovo fattori di ordine politico generale, imperiale, vennero gravemente a pesare sulle condizioni della realtà socio-economica italica. L'instaurazione dell'impero con Augusto significò, in una prospettiva globale, l'emergere delle province e l'inizio del declino dell'Italia, che si era largamente esaurita nell'immenso sforzo al quale le guerre civili del primo secolo a.C. l'avevano sottoposta. L'agricoltura italica nel Centro Sud andò declinando anche per un'accentuata crisi demografica. Il compito gravoso e rischioso di approvvigionare Roma fu sempre più assunto dalle province, in primo luogo dall'Egitto. Le preoccupazioni del governo romano (manifestate in tutte le età da Tiberio a Traiano e riflesse in Tacito) sono dovute soprattutto alla consapevolezza della precarietà di questa dipendenza dall'esterno e della sua perniciosa influenza sull'economia italica. Tentativi di rivitalizzare l'agricoltura italica (mi riferisco sempre all'Italia Meridionale) ebbero scarsi effetti. Il latifondo a cultura estensiva guadagnò terreno: l'economia dei *saltus* si espanse a danno di quella dei *fundi*, che invece si sviluppò nella valle del Po. E' molto importante il rilievo (di B. Ward Perkins) che la soppravvivenza diffusa nell'Italia Settentrionale, sul terreno, della centuriazione romana può, e forse deve, testimoniare la continuità del modo di sfruttamento del suolo, che si era sviluppato proprio a seguito del riordinamento agrario dovuto alla tecnica agrimensoria della centuriazione. Là dove, nel Centro Sud, questa sopravvivenza è venuta meno o è in ogni caso molto meno apparente può indicare un cambiamento di tipi di culture e di utilizzazione del suolo e un ritorno a pratiche anteriori al processo di romanizzazione-municipalizzazione.

Per molto ragioni nelle aree del Centro Sud andò estendendosi la proprietà imperiale, scarsamente produttiva. Venne meno, se pur mai c'era stato, l'incentivo dovuto ad un'espansione del fattore cittadino, che aveva avuto, invece, effetti così positivi nell'Italia Settentrionale. In queste condizioni l'allevamento andò crescendo, con l'ampliarsi delle zone trasformate a pascolo. Lo stesso sistema della transumanza andò largamente accentrandosi nelle mani dell'amministrazione imperiale. La famosa iscrizione di Saepinum (169-172 d.C.) parla di violenti contrasti fra le autorità locali di Bovianum e di Saepinum e gli appaltatori privati delle greggi imperiali. Intervennero i prefetti del pretorio. Un'ipotesi recente ha sostenuto che le greggi danneggiate non fossero soltanto imperiali, ma anche private.[4] In ogni caso non cambia il significato generale del testo, che testimonia come la demanializzazione dei tratturi e il loro uso gratuito, con le garanzie dell'organizzazione statale, non fossero sufficienti a mettere al riparo le greggi da interventi di controllo di magistrati locali, che, da parte loro, avranno avuto buone ragioni per intervenire contro abusi, sconfinamenti, abigeati. Che in queste aree si fosse sviluppata da tempo una forma endemica di brigantaggio è indicato da molte testimonianze e proprio per l'età tardo-antica il divieto dell'uso dei cavalli come misura preventiva nell'Italia Suburbicaria (C.Th.IX 30) conferma ulteriormente la validità dell'equazione pastores-latrones almeno nella visuale del potere.

Val la pena di ricordare che cinque secoli dopo, verso la metà del settimo secolo, quelle stesse aree di Saepinum e Bovianum erano completamente spopolate e i Langobardi di Benevento vi stanziarono nuclei di Bulgari (Paul. Diac. *Hist. Lang.* 5.29).

6. La struttura, complessa e articolata, del sistema della transumanza, messa in essere dall'amministazione romana, era in ogni caso legata alla capacità operativa e all'efficienza dello stesso potere centrale. Il declino, se non addirittura la crisi, dei poteri pubblici dalla fine del secondo secolo d.C. compromise globalmente l'esistenza stessa delle infrastrutture create e amministrate dallo stato e che garantivano in Italia i ritmi della vita sociale e politica. Al decadimento delle vie di communicazione terrestri, delle pratiche di regolamento dei corsi d'acqua, corrispose in molte aree centro-meridionale anche il venir meno dello strutture municipali cittadine. Anche il sistema ufficiale della transumanza deve aver iniziato una fase di diclino, o meglio di *ritorno* verso le condizioni esistenti prima del suo sviluppo, iniziato nel secondo secolo a.C.

Come ho già avuto occasione di sostenere in una relazione[5] non credo che la transumanza sia venuta meno fra il settimo e il dodicesimo secolo.[6] Non si tiene nel dovuto conto la casualità delle nostre conoscenze di quel fenomeno anche nella sua fase classica.

Penso, per converso, che la transumanza sia andata perdendo, più o meno lentamente, la sua complessa organizzazione e il suo scopo fiscale, che poteva reggere soltanto se sostenuta da un autorità politica centrale forte. D'altro canto le direzioni dell'economica italica sono completamente mutate con il terzo secolo d.C. proprio per il venir meno della posizione centrale e imperiale di Roma. Sono sempre meno i ricchi disposti ad investire i loro capitali in Italia. Anche se abbiamo attestazioni sicure del perdurare del sistema della transumanza fra quarto e sesto secolo, la sua ampiezza si andò riducendo. Deve essere invece ritornato, nelle aree appenniniche, a prevalere il tipico sistema silvo-pastorale arcaico. Un indizio del regresso all'allevamento brado, specialmente dei suini, è stato identificato (A. Giardina) nel passaggio alla *caro procina* nell'approvvigionamento di Roma nel quarto secolo. Un regresso dell'economia italica in questo secolo è ammesso, pur nel suo generale ottimismo, anche dal Whittaker. Il quadro generalmente positivo dell'agricoltura nell'età teodoriciana, quale emerge dalle notazioni di Cassiodoro,[7] deve essere valutato tenendo presente la qualità della fonte.

Naturalmente so bene quanta ambiguità vi sia nello stesso concetto di regresso, che tuttavia in questo caso rivestirebbe un valore profondamente diverso rispetto alle teoria che abbiamo discusso all'inizio e riferite al secondo secolo a.C. Wickham,[8] ha messo in chiaro il fondamento ideologico che domina spesso le nostre fonti nell'opposizione di un sistema economico sviluppato, quello agricolo, di contro a quello pastorale, emarginato nell' opinione comune. Tuttavia, anche se riconosciamo la coesistenza e fin l'equilibrio in età romana fra terreni coltivati da un lato e foresta e terreni acquitrinosi o abbandonati dall'altro,[9] è proprio nella rottura di quell'equilibrio e nel prevalere, in certi momenti storici, di un sistema sull'altro che si può e si deve valutare il progresso o il ritorno. Orbene il declino della posizione e della funzione centralizzante di Roma e del suo mercato in età tardo-imperiale, in relazione ai mutamenti generali dell'impero, ha favorito e sollecitato un fase di 'regresso', vale a dire un riemergere dell'economia silvo-pastorale che, come si è detto, è parallela al ritorno a forme di vita non urbanizzate in molto aree del Centro-Sud. Il problema è tanto più importante in quanto ripropone la tematica generale della continuità fra evo antico e alto Medioevo. Come hanno ripetuto i recenti lavori del Wickham, uno degli elementi di questo continuità strutturale è rappresentato proprio dal predominio nelle aree centrali della penisola di terre fiscali, che hanno facilitato il riemergere di forme di sfruttamento collettivo del suolo e quindi di un predominio della pastorizia, non industrializzata. D'altro canto la coincidenza, fra l'età antica e la medievale, del quadro sociale e delle basi economiche della vita associata in alcuna aree tipiche della transumanza antica e moderna (per es. l'area dei Vestini) rappresenta, mi sembra, un forte argomento per sostenere una sostanziale stabilità di condizioni, che ci sono meglio documentate per l'età tardo-repubblicana e per l'Alto Impero.

UNIVERSITÀ DI PAVIA EMILIO GABBA

NOTES

1. C. e E. Tomaselli, 'Appunti sulle vicende delle foreste padane dall'epoca romana ad oggi', *Archivio Botanico e Biogeografico Italiano* 2 (1973) 85-101.

2. G. Traina, *Le Grandi Valli Veronesi in età romana* (1983) 91-4.

3. *Peasants in Peril. Location and Economy in Italy in the Second Century B.C.* (1984).

4. M. Corbier, 'Fiscus and patrimonium: the Saepinum inscription and transhumance in the Abruzzi' *JRS* 73 (1983) 126-31.

5. 'La transumanza nell'Italia romana. Evidenze e problemi. Qualche prospettiva per l'età alto medievale' in *L'uomo di fronte al mondo animale nell'alto Medioevo* (*Settimane di Studio di Spoleto*, 31 (1985)) 373-89.

6. Come talora si suppone, magari in conseguenza dei disastri della guerra grecogotica: Wickham (infra). Tentativi recentissimi di sostenere una quasi completa sparizione del sistema della transumanza per quei secoli non convincono: A. Clementi, 'La transumanza nell' Alto Medioevo', *Bull. Deput. Abruz. Storia Patria* 74 (1984) 31-47.

7. R. Soraci, *Aspetti di storia economica italiana nell' età di Cassiodoro* (1974).

8. C. Wickham, 'Pastoralism and underdevelopment in the early Middle Ages' in *Settimane di Studio di Spoleto* 31 (1985) (n.5 supra) 401-51; per Italia spec. 412-13.

9. G. Traina, 'Antico e moderno nella storia delle bonifiche italiane' *Studi Storici* 2 (1985) 431-6.

VIEHZUCHT, ALPWIRTSCHAFT, TRANSHUMANZ: BEMERKUNGEN ZU PROBLEMEN DER WIRTSCHAFT IN DER SCHWEIZ ZUR RÖMISCHEN ZEIT

Im Mittelpunkt der folgenden wirtschaftsgeschichtlichen Überlegungen steht das Problem der Viehzucht in Verbindung mit Alpwirtschaft und Transhumanz. Es ist dabei offensichtlich, dass dieser Problembereich nicht generell am Beispiel der Schweiz in römischer Zeit untersucht werden kann, da hier verschiedene Wirtschaftsräume voneinander getrennt werden müssen, nämlich Alpen und Teile der Voralpen einerseits und das Schweizerische Mittelland mit dem besiedelten Raum des Jura anderseits, wobei nur die letzteren Gebiete von der römischen Wirtschaftsform der *villa* in verschiedener Stärke erfasst worden sind. Auch dürfte es vorzuziehen sein, die besonderen Bedingungen des Wallis gesondert zu untersuchen und in der vorliegenden Analyse wegzulassen.[1]

Aus diesem Grunde wird zuerst der Problemkreis der Viehzucht in den Alpen und Voralpen behandelt, wobei hier nochmals die Quellen zur alpinen Wirtschaft vorgelegt werden; bekanntlich spielt die Viehwirtschaft in den Alpen und in den inneralpinen Tälern eine traditionell grosse Rolle und steht deshalb auch hier im Vordergrund. Darauf werden anhand des Begriffes der Transhumanz Verbindungen zum Schweizerischen Mittelland geprüft und schliesslich einige Fragen zur Viehzucht im Mittelland aufgeworfen, wo für die genauere Erfassung der Abläufe noch viele Fragen offen stehen. Einige allgemeinere historische Bemerkungen sollen den Schluss bilden. Freilich handelt es sich bei allen folgenden Ausführungen nicht um die Darlegung ausführlicher eigener Forschungen, sondern nur um den Versuch, eine vorläufige Standortbestimmung zu geben, wo mehr Fragen gestellt als beantwortet werden können.

1. Wirtschaftsformen in den Alpen: Forschungsansätze

Das Forschungsgebiet der antiken, prähistorischen und römerzeitlichen, alpinen Wirtschaftsformen[2] steht bekanntlich im Schnittpunkt ganz verschiedener Forschungsansätze. Da ist an erster Stelle die traditionelle und immer noch wichtige, da sinngebende, Interpretation der antiken Primär- und Sekundärnachrichten zu nennen; dann folgt – noch bedeutender, da man sich trotz der römischen und damit historischen Zeit für dieses Thema gewissermassen in beinahe schriftloser, prähistorischer Epoche befindet – die Archäologie, die neben der Darlegung des Fundmateriales eines einzelnen Grabungsplatzes immer mehr auch versucht, zu Gesamtsichten im Sinne der Siedlungsarchäologie zu kommen,[3] wobei unter Einbezug des biologischen Umfeldes mit Pollenanalyse und Analyse der erhaltenen Tierknochenfunde auch neue historische Fragen an das Material gestellt werden.[4] Ein weiterer, fruchtbarer Ansatzpunkt, der für die Prähistorie bereits mit den oben genannten Ansätzen kombiniert, aber für die römische Zeit doch erst in Umrissen angewendet wird, lässt sich mit dem Begriff Klimageschichte und ihrer Auswirkung auf Mensch und Wirtschaft umschreiben,[5] wo insbesondere aufgrund von Gletscheruntersuchungen und Dendroanalysen in Kombination mit weiteren, für die Neuzeit auch literarischen, Nachrichten versucht wird, das Klima und seine Auswirkungen auf das Leben und die Wirtschaft eines ausgewählten geographischen Bereiches in einer bestimmten Epoche genauer zu fassen.[6] Zwei speziellere Fragestellungen lassen sich noch zusätzlich anführen. Eine erste beschäftigt sich mit dem Zeitpunkt der Eroberung der Alpen durch den Menschen in der Prähistorie und damit verbunden mit dem Beginn und den besonderen Formen der Alpwirtschaft;[7] eine zweite sucht das Problem der Transhumanz, in die vorläufig die Alpwirtschaft eingeschlossen wird, in diachronischen Untersuchungen in verschiedenen geographischen Räumen anzugehen.[8]

2. Literarische und epigraphische Zeugnisse zur Wirtschaft in den Alpen

Da die wenigen literarischen Zeugnisse über die Wirtschaft in den Alpen schon oft dargestellt worden sind,[9] können die Ausführungen recht kurz gefasst werden; dagegen scheinen die Inschriften offenbar nicht so bekannt zu sein, so dass sie eine ausführlichere Behandlung erfordern. Die griechischen und römischen Schriftsteller haben sich bekanntlich erst spät und nur an ganz wenigen Stellen mit Bewohnern der Alpen und der nördlichen Voralpen auseinandergesetzt.[10] Was die Wirtschaftsbedingungen betrifft, so ist ihr Blickwinkel sehr beschränkt, äussern sie sich doch fast nur über Erzeugnisse, die nach Italien importiert worden sind, während umgekehrt Exportprodukte in die Alpengebiete oder Nachrichten über inneralpine Wirtschaft in den literarischen Quellen nicht in Erscheinung treten.

Das Hauptzeugnis für die alpine Wirtschaft gibt Strabo, der auf Polybios oder Poseidonios zurückgehen mag,[11] womit die Stelle einem etwas älteren Zustand entsprechen, aber in den Grundzügen wohl auch noch auf die römische Kaiserzeit zutreffen dürfte.

> Strabo 4.6.9: 'In dem ganzen Bergland der Alpen gibt es auch hügelige Gegenden, die gut bebaut werden können, und trefflich angebaute Täler, der grössere Teil aber und besonders um die Berggipfel herum, wo auch die Räuber sassen, ist wegen des Frostes und der Rauheit des Landes unergiebig und unfruchtbar. Bei dem Mangel an Nahrungsmitteln und anderen Dingen

schonten sie bisweilen die Bewohner der Ebene, um Lieferanten zu haben. Dafür aber gaben sie Harz, Pech, Kienholz, Wachs, Käse und Honig, denn davon hatten sie in Menge.[12]

Diesem ethnographischen Hinweis lässt sich entnehmen, dass der Handel und die Handelsbeziehungen vorwiegend zwischen den Bündner Alpen, beziehungsweise den Südtiroler Alpen, Dolomiten, Trientiner Alpen[13] und der Poebene und ihrer nördlichen Randzone stattfand. Der Warenaustausch lässt 'auf Waldwirtschaft und Alpwirtschaft nebst Bienenzucht oder Gewinnung von wildem Honig' schliessen.[14] Diese Grundzüge werden, ohne dass sehr viele weitere Einzelheiten hinzugefügt würden, von Plinius in seiner Naturgeschichte bestätigt. An erster Stelle sind hier die Notizen zur Alpwirtschaft und Viehzucht zu erwähnen. Bekannt ist seine Bemerkung, dass die Alpenkühe am meisten Milch gäben[15] und dass in Rom Alpenkäse,[16] wohl Hartkäse,[17] sehr geschätzt wurde. Hingegen ist es wohl kaum zulässig, eine zweifelsfrei überlieferte Stelle bei Columella dahin zu emendieren, dass die Römer für die Aufzucht der Kälber, die in der Antike mit Ammenkühen vor sich ging, milchreiche Alpenkühe importiert und verwendet hätten.[18] Der Text Columellas spricht von *vaccae Altinae*, von Kühen um die Stadt Altinum.[19]

Zweitens weist der Text des Plinius auf die Bedeutung der *pabula*, wohl mit '(Alp)weiden' zu übersetzen, hin, allem Anschein nach die einzige Stelle, die im Zusammenhang mit der Wirtschaft des Alpengebietes auf die Qualität der Weiden eingeht.[20] Im oben zitierten Text Strabos wird auch deutlich, dass, wo es möglich war, Getreide angebaut wurde, was Plinius mit einem Hinweis auf eine besondere Weizensorte in seiner Naturgeschichte bestätigt.[21] Eine Reihe anderer Stellen bezeugt den Waldreichtum und die hohe Qualität der Nadelbäume, insbesondere der Lärchen, die in Rom als Bauholz verwendet wurden.[22] Die Exportprodukte von Honig und Wachs fügen sich in das Gesamtbild ein.[23] Die Metalle und Gesteine als weitere, wirtschaftlich interessante Rohstoffe sollen hier nur im Vorbeigehen erwähnt werden.[24] Zum Schluss sei noch eine literarische Stelle, nämlich einige Verse aus dem Gedicht des Claudian, *De bello Pollentino*, herangezogen, in denen die Überquerung eines Passes mitten im Winter durch Stilicho geschildert wird. Festzuhalten ist daraus die Nennung von Hirten- oder Alphütten, in denen Stilicho nächtigte, und zudem der Umstand, dass Wein dort offenbar nicht üblich und Brot offenbar nur wenig gegessen wurde;[25] freilich ist es nicht sicher, ob es sich bei dieser Schilderung nicht eher um einen literarischen Topos handelt.[26]

Neben diesen spärlichen Reflexen der alpinen Wirtschaft in den Schriftstellerberichten der römischen Kaiserzeit sind die Inschriften als Primärzeugnisse an die erste Stelle zu setzen. Leider gibt es in diesem geographischen Bereich nicht die grossen und interessanten Inschriften über Weidekonflikte wie aus anderen Gegenden.[27] Immerhin spiegeln die wenigen Weihinschriften aus Graubünden doch Aspekte des Wirtschaftslebens wider.[28] So sind hier die vier kleinen, aus Lavezstein angefertigten Weihaltärchen zu nennen, die im Kiessand des Silsersees bei Sils-Baselgia gefunden wurden.[29]

AE 1966, 272: *Dianae/ Tertius/ Valerii/ u(otum) s(oluit) l(ibens) m(erito)*
AE 1966, 273: *Mercur[io/ Tertius/ Valerii/ u(otum) s(oluit) l(ibens) m(erito)]*
AE 1966, 274: *Pastoribus/ Tertius/ Valerii/ u(otum) s(oluit) l(ibens) m(erito)*
AE 1966, 275: *Siluano/ Tertius/ Valerii/ u(otum) s(oluit) l(ibens) m(erito)*

Die Altäre weisen eine parallele Textgestaltung auf, sind von einem *Tertius Valerii* gestiftet und an die Gottheiten Diana, Silvanus, Merkur und die *pastores*, die Hirtengötter, gerichtet. Das Material der vier Weihaltärchen ist der bereits von Plinius als Lokalstein erwähnte Lavez- oder Speckstein, aus dem auch die Altärchen an *Mercurius Cissonius* aus dem Bergell[30] und interessanterweise ein in Augst gefundener kleiner Altar[31] bestehen und der offenbar im näheren Umkreis, etwa im Raum von Chiavenna oder im Bergell, verarbeitet wurde. Wie die Fundstelle zeigt, sind die Inschriftensteine über den Maloja ins Engadin transportiert worden und dort offenbar beim Transport verloren gegangen.[32] Damit sind sie bereits durch das Material und den Ort des Verlustes seltene Zeugnisse des Lokalverkehrs in diesem schlecht bezeugten Gebiet. Die Vermutung liegt nahe, dass sie für diesen geographischen Raum bestimmt gewesen waren, wobei die Datierung der Altäre ans Ende des 2. oder in den Anfang des 3.Jh. gesetzt werden kann. Der Stifter und Auftraggeber, ein *Tertius Valerii*, darf jedenfalls als Peregriner im Sinne von 'Tertius, Sohn des Valerius' interpretiert werden, dessen Namensform einheimischer Tradition entspricht.[33] Die angerufenen Götter sind zweifellos lokale Gottheiten; wenn man auch nicht weiss, ob sie in dieser Zusammenstellung verehrt wurden, so beleuchten sie doch Aspekte des dortigen Wirtschaftslebens: Merkur als der Gott des Handels und Verkehrs, der in einer Sonderform auch auf den Weihaltären aus Promontogno verehrt wird, steht in engstem Zusammenhang mit der Passroute; auf welche Weise die Pässe seit frühesten Zeiten von den Bewohnern der Alpentäler mit Säumer-, Transport- und Begleit- und Umladediensten ausgewertet und bewirtschaftet wurden, hat Denis van Berchem für die Walliser Übergänge dargelegt.[34] Man darf zweifellos die gleichen Verhältnisse auch für die Bündner Pässe voraussetzen. Bei der Weihung an Diana und Silvanus, hinter welchen sich trotz des Fehlens des sonst üblichen Zusatzes *deus* oder *dea* einheimische Naturgottheiten verbergen dürften, liegt der Zusammenhang zwischen Wald und Jagd auf der Hand. Eine Weihung an die gleichen Götter ist aus Zürich/Turicum, gesetzt von den Bärenjägern, belegt.[35]

Einmalig und bis jetzt recht wenig gewürdigt ist dagegen die Nennung der *pastores*, der Hirtengötter. Ernst Meyer und Hans Georg Pflaum machen mit aller Zurückhaltung auch hier auf eine mögliche Parallele, eine Weihinschrift in Gedichtform aus Italien, aufmerksam, wo Silvanus als *sanctissime pastor* angesprochen wird.[36] Freilich dürfte diese Parallele doch etwas weit hergeholt sein; in der Inschrift aus Sils-Baselgia handelt es sich wohl um echte, einheimische Hirtengötter, die sonst bisher nirgends in Erscheinung getreten sind. Damit wird aber diese unscheinbare kleine Inschrift zu einem wertvollen Zeugnis, belegt sie doch von epigraphischer Seite her wohl für das Engadin, jedenfalls aber für den grösseren geographischen Raum, Viehzucht und gegebenenfalls Alpwirtschaft in irgendeiner Form.

Die beiden weiteren, hier auswertbaren epigraphischen Zeugnisse, zwei Weihinschriften aus dem Bergell,[37] beziehen sich mit der Nennung von *Mercurius Cissonius* auf den Passverkehr über den Septimer und gehören also nicht mehr zum hier behandelten Thema der Viehzucht.[38]

3. Zu den archäologischen Zeugnissen

Verschiedene Forscher haben bereits festgestellt, dass die Materialgrundlage noch zu schmal ist, um allgemeinere Aussagen über die Besiedlung der Alpentäler in römischer Zeit und ihre Wirtschaftsstruktur wagen zu können.[39] Bisher sind nämlich keine Funde gemacht worden, die etwa die Alpwirtschaft in römischer Zeit zweifelsfrei belegen würden, und nur sehr wenige, die die Besiedlung der Alpentäler in höheren Lagen nachweisen. Die Forschung hat sich bis jetzt notgedrungen auf den Nachweis von Gutshöfen (*villae*) beschränken müssen, deren Grundmauern eher zufällig bei Bauvorhaben angeschnitten wurden; immerhin haben die Grabungen auf der Cadra von Riom-Parsonz bereits interessante Ergebnisse erzielt.[40] Dann stand die Untersuchung der Passstrassen im Zentrum, wo das Trassee und die Spuren des römischen Ausbaues diskutiert werden.[41] Weitere Fundstellen sind noch eher selten.[42]

Diese Forschungssituation kann nicht recht befriedigen angesichts des Umstandes, dass die intensive Urgeschichtsforschung, vor allem die Erforschung der Bronzezeit, die Inbesitznahme der Alpen durch den Menschen nachweisen konnte. Wenn auch die Diskussion über die vormittelalterliche Alpwirtschaft noch im Gang ist, so haben doch die Arbeiten von René Wyss für die Schweizer Gebiete und jene von R. Pittioni und anderen für die Ostalpen wohl eindeutig zeigen können, dass der Mensch der Bronzezeit gewissermassen die Alpen erobert hatte, und dass die menschliche Besiedlung ganzjahreszeitlich oder temporär höher gelegene Regionen um 1000 m/M (Hügel- oder Höhensiedlungen in Talschaften) bis jene von 1800m/M-2100m/M (echte Höhensiedlungen und Siedlungen in Hochlage) erfasst hatte.[43] Ohne hier nun auf die verschiedenen Motive dieser Ausdehnung des menschlichen Bereichs, die vor allem durch die Suche nach Metallen bestimmt war, eingehen zu wollen, so kann man doch feststellen, dass damit die Vorbedingung für die Begründung der Alpwirtschaft erfüllt wurde, nämlich das Heranrücken der Talsiedlung an die Bergzone, von wo aus die Weidegebiete leicht erreichbar waren.[44] Allerdings kann, wie auch Ludwig Pauli einräumt, die Existenz einer Alpwirtschaft in der Frühzeit (prähistorische und römische Zeit) immer noch nur indirekt erschlossen und nicht eindeutig bewiesen werden, wenn sich auch die Indizien dafür immer mehr verstärken. So sind die Höhenfunde (Beile und Speerspitzen) nicht lediglich als Jagdverluste und damit als Zeugnis der zufälligen und nur zeitweisen Ortsbegehung zu werten, sondern aufgrund ihrer Lage in der Nähe von auch heute noch benutzten Alpweiden versteht sie René Wyss, dem sich weitere Forscher anschliessen, als Opfergaben von Hirten und Sennen.[45] Zu den nicht zu vernachlässigenden, wenn auch schwierig zu interpretierenden Indizien gehören schliesslich auch die von sprachwissenschaftlicher Seite hervorgehobenen 'Alpenwörter', die oft einer sehr frühen Sprachstufe zuzuordnen sind.[46]

Die Schwierigkeit für die Beurteilung der Situation zur römischen Zeit besteht nun aber darin, dass die Funde mit der Bronzezeit abbrechen, was mit der Klimaverschlechterung am Ende der Bronzezeit in Zusammenhang gebracht wird; nach anderer Ansicht entspricht der Rückgang der Funde einer Änderung der Sitte der Weihegaben und bedeutet keinen Rückgang der Höhensiedlungen.[47] Aus Rückschlüssen aus den Klimaschwankungen des Mittelalters und der Neuzeit, die sich im Auf und Ab der Alpwirtschaft spiegeln, dürfte die Erklärung des Klimasturzes doch die wahrscheinlichere sein, wenn auch unbestritten eine Änderung der Weihgabensitte zu beobachten ist. Für die römische Zeit ist sich jedenfalls die Klimaforschung einig, dass sich von 300 v.Chr. bis 400 n.Chr. die Gletscher zurückgezogen hätten und eine spürbare Klimaverbesserung eingetreten sei, die übrigens auch literarisch für Italien überliefert ist.[48]

Deshalb wird man die Frage stellen dürfen, ob der Mensch nicht auch damals in beschränktem Masse die Schwankungen des Klimas, wie auch im Mittelalter und in der frühen Neuzeit, nachvollzogen und in der römischen Kaiserzeit wiederum die höher gelegenen Siedlungen aufgesucht sowie die Alpweiden wieder benutzt habe, wobei die Benutzung der natürlichen Alpweiden vorausgesetzt wird.[49] Zeugnisse für diese Hypothese liegen offenbar bis jetzt nicht vor. Ja sogar, Zeugnisse, die man in der früheren Forschung für solche Belege der römerzeitlichen Alpwirtschaft gehalten hatte, nämlich die sog. 'Heidenhüttli', Reste abgegangener Hütten,[50] müssen differenziert beurteilt werden. Die mustergültige Untersuchung einer solchen Fundstelle, Bergeten ob Braunwald(GL), hat nämlich ergeben, dass die Zeitstellung dieser temporären Siedlung zwischen dem 12. und 15.Jh. anzusetzen ist.[51] Freilich dürfen diese Schlussfolgerungen nicht verallgemeinert werden; aber ein Nachweis römischer Alphütten und damit auch Alpwirtschaft ist gerade wegen der ärmlichen Ausstattung solcher Hütten und auch wegen der immer neuen Belegung der möglichen Fundplätze ausserhalb der späteren Rodungsgebiete sehr schwierig zu führen. Somit bleibt die Annahme einer Alpwirtschaft in römischer Zeit, die durch die oben genannten literarischen und epigraphischen Belege nahegelegt wird, vorerst noch Hypothese.

4. Alpwirtschaft und Transhumanz

In diese Überlegungen müssen nun auch die Ergebnisse der geographischen Forschung über die begriffliche Differenzierung zwischen Alpwirtschaft und Transhumanz sowie jene der historischen geographischen Forschung über die Transhumanz in der Schweiz einbezogen werden. Wenn auch noch oft von der Alpwirtschaft als von Transhumanz gesprochen wird, so hat doch Hans Boesch, dem Stanislaw Berezowski, Dorothea Zöbl und andere gefolgt sind, den Nachweis erbracht, dass Alpwirtschaft von Transhumanz im engeren Sinn getrennt werden muss und eine eigene Wirtschaftsform darstellt.[52] Transhumanz stellt nach Hans Boesch eine lose Kombination von Ackerbau und fester Siedlung einerseits mit Viehhaltung anderseits dar, wobei der Besitzer der Herden mit seiner Familie im Dorf lebt, Acker und Rebberg bestellt, allenfalls einen nichtlandwirtschaftlichen Beruf ausübt, und die Obhut über die Herden, vor allem von Schafherden, einem Hirten übergibt. Dieser zieht der Jahreszeit entsprechend den Weiden nach, sucht im Sommer die Bergregionen, im Winter das Tiefland auf.[53] Von dieser Beobachtung ausgehend dürfte es einleuchten, dass die Transhumanz in den Schweizer Alpen nicht möglich ist, da hier die Voraussetzungen für einen winterlichen Weidegang fehlen, was ein wesentliches Unterscheidungsmerkmal zur Transhumanz darstellt. Deshalb definieren Hans Boesch und Josef Birkenhauer die Alpwirtschaft als eine eigenständige Wirtschaftsform,[54] bei der das Vieh im Winter im Stall durchgefüttert wird – was einen schwierigen Futterengpass bedeuten kann – und bei der die Distanz zwischen Sommer-und Winterstandort kleiner ist als bei der Transhumanz. Weiter stellen bei dieser Wirtschaftsform Tal- und Bergbetrieb ein untrennbares Ganzes dar, indem das Vieh einerseits teilweise aus der Futterfläche des Talgebietes ernährt wird, anderseits die Maiensässe (Temporärsiedlungen) und Alpweiden als Futterlieferanten in die Talwirtschaft einbezogen werden. Die beiden Wirtschaftsformen, Transhumanz und Alpwirtschaft, sind also deutlich getrennt und nach Hans Boesch ist die Alpwirtschaft die in der Schweiz vorherrschende Form. Es dürfte einleuchten, dass diese Art der Alpwirtschaft auch für prähistorische und römische Zeit möglich ist und angenommen werden darf, auch wenn sie noch kaum belegt werden kann und auch wenn zweifellos zwischen verschiedenen Stufen der Inbesitznahme von hohen und höher gelegenen Maiensässen und Alpweiden unterschieden werden muss. Die neueste Untersuchung zur Bronze- und Eisenzeit des Unterengadin von Lotti Stauffer-Isenring zeigt denn auch, dass dort eine Selbstversorgungswirtschaft mit kombiniertem Ackerbau und Viehzucht anzunehmen ist, wobei die Ausdehnung auf eine mehr oder weniger umfangreiche Alpwirtschaft erwogen wird.[55]

Zwei Einschränkungen müssen noch angebracht werden, auf die Dorothea Zöbl und Rolf Gubler-Gross aufgrund ihrer geographisch-historischen Studien hinweisen.[56] Der Überschuss an Magerweiden in den Alpen ist bis zum Ersten

Weltkrieg als Sommerweide für Schafherden aus dem Mittelmeergebiet im weitesten Sinn, aus Oberitalien (Bergamasker Schafe), verpachtet worden, womit sich die Transhumanz von Süden her bis in den Bereich der Schweizer Alpen erstreckte. Umgekehrt ist auch ein allfälliger Viehüberschuss aus den Bündner Tälern im Herbst in der Lombardei verkauft worden. Anderseits hat sich nach den Untersuchungsergebnissen von Rolf Gubler-Gross in der Schweiz eine moderne zentralalpine Transhumanz entwickelt, die ebenfalls vom Überschuss an Schafweiden in den Alpen im Sommer ausgeht.[57] Hier wohnt der Besitzer der Herden, meist ein Händler oder Metzger, im Mittelland oder im Voralpengebiet, sömmert die Herde in den Alpen, lässt sie im Winter in schneearmen Regionen des Mittellandes weiden und sorgt mit Pacht von Weideland oder Kauf von Futter für die Ernährung in der futterarmen Zeit des Frühlings und Herbstes. Dieses recht differenzierte System hat sich offenbar erst nach dem politisch bedingten Rückgang der Transhumanz der Bergamasker Schafe im Ersten Weltkrieg entwickelt und ist somit modern. Eine Uebertragung auf antike Verhältnisse scheint deshalb wohl ausgeschlossen zu sein. Dagegen mag die erstere Form der Transhumanz von Süden her auf sehr alte, wohl auch antike Zeit zurückgehen; auf Verbindungen vom Nordrand der Poebene in das Alpengebiet ist oben aufgrund literarischen Quellen hingewiesen worden.

Die Schlussfolgerungen, die sich aus diesen Studien für die Schweiz in römischer Zeit ziehen lassen, können deshalb etwa wie folgt umschrieben werden: Offenbar müssen für die Antike recht kleinräumige Verhältnisse angenommen werden: Einerseits ist Alpwirtschaft zu vermuten, die aber wahrscheinlich in kleinen, überschaubaren Bereichen und Talschaften betrieben wurde und die fast keine Nachrichten hinterlassen hat; anderseits bildete die Viehzucht einen Teilbereich der Wirtschaft in den *uillae* des Mittellandes und des Jura, welche von jener der Alpen getrennt werden muss.

5. *Viehhaltung im Schweizerischen Mittelland*

Zur Viehzucht in den *villae* des Schweizerischen Mittellandes seien nur allgemeine Bemerkungen angefügt, da auch hier die Erforschung der Oekonomieteile der Gutshöfe noch viele Lücken aufweist.[58] Die wenige materielle Hinterlassenschaft besteht aus Viehglocken, aus einer Kastrierzange, aus Utensilien zur Milchverarbeitung und bezeugt somit Viehzucht und Viehmast;[59] die Tierknochenfunde belegen Rinder, Schweine, Schafe und Ziegen, wobei die ersten beiden Tierarten weit überwiegen.[60] Aus der Anlage der Bauten und Grundmauern schliesst Walter Drack im Falle der *uilla* von Seeb auf offene Viehpferche, zudem könnte die *uilla* bei Kloten einen Kühlkeller für Milchprodukte aufgewiesen haben.[61] Die Beweise reichen also aus, um Viehzucht in den Gutshöfen zu belegen.

In der Umgebung grösserer Agglomerationen, etwa von Augusta Raurica, scheinen die *uillae*-Besitzer auch bewusst nach römischem Vorbild für die Stadt und die Essgewohnheiten der reicheren Bewohner produziert zu haben, hat doch Elisabeth Schmid in einer Grossküche von Augst Spanferkelknochen nachweisen können.[62] Ebenso müssen gerade für Augusta Raurica die Räucherkammern erwähnt werden, die als Stätten der Fleischverarbeitung für den Export interpretiert werden; auch wenn die letztere Annahme nicht gesichert ist, so lassen diese Räucherkammern doch auf eine entwickelte Viehzucht schliessen.[63]

Damit stellt sich die Frage nach den Formen der Viehhaltung in diesen Gebieten. Aus Analogieschlüssen[64] wird man vermuten dürfen, dass wohl eine Fruchtwechselwirtschaft bestanden hat, wobei die Reihenfolgen noch nicht bekannt sind. Das Vieh weidete offenbar auf abgeernteten Feldern und auf gemähter Brache, wo die Gutsbesitzer möglicherweise Klee eingesät haben, um gutes Heu und nachher gute Weide zu erhalten. Fettwiesen wurden jedenfalls gemäht, ebenfalls Magerwiesen, die desgleichen zur Tierweide dienten. Weitere Erkenntnisse, insbesondere über das Verhältnis von Ackerfläche zu Grünfläche sind sehr schwer zu erzielen und bisher nicht möglich gewesen. Der Anschluss an Forschungsergebnisse mittelalterlicher Studien könnte weitere Erklärungsvorschläge anbieten. Die Untersuchungen von Nicolas Morard zur Alpwirtschaft in den Freiburger Voralpen haben ergeben,[65] dass zudem auch Wälder bis zu 1000m/M die Nahrungsgrundlage für das Vieh dargestellt haben. In den Vogesen sind anderseits archäologische Spuren von sog. Viehwegen entdeckt worden, die auf eine Viehhaltung in grösserem Umfang in *saltus* hindeuten;[66] ob solche Spuren auch in der Schweiz, etwa im Jura, je gefunden wurden und ob sie überhaupt vorauszusetzen sind, ist unbekannt. Immerhin bieten sich die Voralpen als weiterer, in diesem Zusammenhang wichtiger Wirtschaftsraum an, der durch die jüngsten Ausgrabungen in der Forschung immer mehr an Bedeutung gewinnt.[67] Allerdings dürfen auch die Dimensionen der Wirtschaftsform des Gutshofes nicht überschätzt werden, da die Untersuchungen der *uillae* im Schweizerischen Mittelland bisher kaum ganz grosse Anlagen zutage gefördert haben;[68] dieser Befund stimmt mit den Ergebnissen der Erforschung der Sozialstruktur überein, die im Gebiet der Helvetier und der übrigen Stämme ebenfalls mit kleineren Verhältnissen rechnet.[69] So ist in diesem vielgestaltigen Wirtschaftsraum des Mittellandes, der Voralpen und der Alpen die Viehzucht gesichert, wobei die angeschnittenen Problemkreise noch der Vertiefung bedürfen.

UNIVERSITÄT BERN REGULA FREI-STOLBA

NOTES

1. Zum Wallis liegt eine reiche Literatur vor, s. den Überblick über die römische Zeit von P. Ducrey, 'Etat de recherche sur le Valais romain', *Vallesia* 33 (1978) 17-30 und zur hier interessierenden Frage

der Romanisierung der Seitentäler F. Wiblé, 'Considérations sur la romanisation des vallées latérales en Valais', *Bulletin d'études préhistoriques alpines* 15 (1983) 189-93 (mit Lit.).

2. S. vor allem *Histoire et civilisations des Alpes*, hsg.v. P. Guichonnet (1980) mit dem Beitrag von D. van Berchem für die römische Zeit (dieser Beitrag jetzt auch in: D. van Berchem, *Les routes et l'histoire* (1982) 185-217, darnach zitiert) und L. Pauli, *Die Alpen in Frühzeit und Mittelalter* (1980) für wirtschaftliche Aspekte bes. 268-88; dazu 'Geschichte der Alpen in neuer Sicht', Schweizer Historikertag 19.Mai 1979, *SZG* 29 (1979) 3-300, mit den hier wichtigen Aufsätzen von P. Ducrey, K. Aerni, M.-R. Sauter und W. Meyer. S. ebenfalls 'La comunità alpina nell'antichità, *Atti del Centro di studi e documentazione sull'Italia romana*, 7 (1975/76). Kurz auch G.Th. Schwarz, 'Die Römer im Bergland der Schweiz', *Germania Romana III.* Römisches Leben auf germanischem Boden (1970) 110-17; F. Staehelin, *Die Schweiz in römischer Zeit*, 3.Aufl. (1948) bes. 389 ff., 424 ff. Zum besonderen, hier nicht behandelten Problem der Wirtschaftsformen in den Südostalpen s. P. Garnsey, 'Mountain economies in southern Europe', *Itinera* 5/6 (1986) 7-29, reproduziert unten, 186–209.

3. L. Stauffer-Isenring, *Die Siedlungsreste von Scuol-Munt Baselgia (Unterengadin GR). Ein Beitrag zur inneralpinen Bronze-und Eisenzeit*, Antiqua 9 (1983); die Autorin geht in ihrer Untersuchung eines geschlossenen geographischen Raumes von ähnlichen Fragestellungen aus, wie sie für den gleichen geographischen Bereich, aber für eine spätere Epoche, formuliert worden sind und bezieht die Ergebnisse ein von Jon Mathieu, *Eine Region am Rand: Das Unterengadin 1650-1800.* Lizentiatsarbeit Universität Bern 1980 (1982). Zur Publikation des Materiales des wichtigen Fundplatzes von Riom-Parsonz (Kr. Oberhalbstein, Bez. Albula) s.J. Rageth, 'Römische und prähistorische Funde von Riom', *BMbl* 3/4 (1979) 49 ff.; ders., 'Römische Funde von Riom', *HA* 8 (1977) 74-7; dazu die Berichte *JbSGU* 59 (1976) S. 265 f.; *JbSGU* 60 (1977) 143 f.; *JbSGU* 64 (1981) 248-52; *JbGSU* 65 (1982) 205-8; *JbSGU* 66 (1983) 296-9; *JbSGU* 67 (1984) 223-5 sowie kurz in *ArchS* 5 (1982) 137-40 und 'Die römischen Schmiedegruben von Riom GR', *ArchS* 5 (1982) 202-8. Zu Graubünden s. ausserdem A. Zürcher, 'Urgeschichtliche Fundstellen Graubündens', *SRMC* 27 (1982); H. Conrad, *Schriften zur urgeschichtlichen und römischen Besiedlung des Engadins*, mit einem Beitrag von S. Nauli (1981); B. Overbeck, *Geschichte des Alpenrheintales in römischer Zeit*, 2 Bde (1978-1982).

4. S. als ein Beispiel die Beiträge bei Stauffer-Isenring (Anm. 3) von B. Kaufmann über die Skelettreste, M. Hopf über die Getreide und Hülsenfrüchte, von W. Schoch über die Holzkohlenreste, von H. Zoller über die naturräumlichen Voraussetzungen; s. ebenfalls bei der Untersuchung des *uicus* von Oberwinterthur den Beitrag von Chr. Jacquat, 'Römerzeitliche Pflanzenfunde aus Oberwinterthur', in: *Beiträge zum römischen Vitudurum-Oberwinterthur* 2 (1986) 241-64.

5. Zur Beziehung zwischen Klima und Geschichte s. neben dem Pionierwerk von E. Le Roy Ladurie, *Histoire du climat depuis l'an mil* (1967), dann engl., *Times of Feast, Times of Famine: a History of Climate since the year 1000* (1971), H.H. Lamb, *Climate, History and the Modern World* (1982); ders., *Climate: Present, Past and Future* – Vol. 1: Fundamentals and Climate Now (1972); Vol. 2: Climatic History and the Future (1977).

6. Die Klimageschichte der Neuzeit ist erstmals auf die Schweiz übertragen worden von Chr. Pfister, *Bevölkerung, Klima und Agrarmodernisierung 1525-1860* 2 Bde (1984). Für die römische Zeit liegen für diese Fragestellung (Kombination von literarischen Nachrichten und naturwissenschaftlichen Erkenntnissen) erst wenige Bemerkungen vor, so etwa Lamb 1977 (Anm. 5) 424 und Lamb 1982 (Anm. 5) 148 ff. mit allgemeinen Hinweisen zu Livius und Saserna bei Columella, die sich auf Italien beziehen. Zur Problematik der Überlieferung s. nun R. Frei-Stolba, 'Klimadaten aus der römischen

Republik', *MH* 44 (1987) 101-17. Zur Dendroklimatologie s.M.K. Hughes, P.M. Kelly u.a., *Climate from Tree Rings* (1982) und die klare, populäre Einführung von F.H. Schweingruber, *Der Jahrring, Standort, Methodik, Zeit und Klima in der Dendrochronologie* (1983). Die naturwissenschaftlichen Erkenntnisse über die Gletscherbewegungen umfassend dargestellt von R. Hantke, *Eiszeitalter. Die jüngste Erdgeschichte der Schweiz und ihrer Nachbargebiete*, 3 Bde (1978-1983) und F. Röthlisberger, *10'000 Jahre Gletschergeschichte der Erde* (1986).

7. S. die Forschungsübersicht von N. Grass unter Mitwirkung von F. Maier-Böttcher, 'Die Almwirtschaft in der Urzeit und im Mittelalter', in *Untersuchungen zur eisenzeitlichen und frühmittelalterlichen Flur in Mitteleuropa und ihrer Nutzung*, hsg. v.H. Beck, D. Denecke und H. Jankuhn, *Abhandl. d. Akad. d. Wiss. Göttingen phil.-hist. Kl. 3*, 115/116 (1979/80) 229-86; dazu *Reallexikon der German. Altertumskunde* 2. Aufl. (1978) Bd. I, 181-189, s.v. 'Alm(Alp)wirtschaft', R. Pittioni, W. König, E. Ploss. Wesentlich auch für die historische Forschung ist nun die moderne Gesamtbearbeitung von J. Birkenhauer, *Die Alpen* (1980).

8. D. Zöbl, *Die Transhumanz (Wanderschafhaltung) der europäischen Mittelmeerländer im Mittelalter in historischer, geographischer und volkskundlicher Sicht*, Berliner Geograph. Studien, B. 10 (1982); für Mitteleuropa in prähistorischer Zeit s. bes. W. Dehn, 'Einige Bemerkungen zu Gesellschaft und Wirtschaft der Späthallstattzeit; "Transhumance" in der westlichen Späthallstattkultur?', in: *Historische Forschungen für Walter Schlesinger*, hsg. v.H. Beumann (1974) 1-18.

9. S. zuletzt Van Berchem (Anm. 2), 208; die Belege gesammelt und übersetzt bei E. Howald/E. Meyer, *Die römische Schweiz* (1940).

10. S. dazu E. Meyer, 'Über die Kenntnisse des Altertums von der Schweiz in vorrömischer Zeit', *Festschrift Hans Nabholz* (1934) 3-21; Van Berchem (Anm. 2) 185-90.

11. Polybios nach F. Lasserre, *Strabon, Géographie*, tome II (1966) 107; Van Berchem (Anm. 2) 208 führt die Vorlage auf Poseidonios zurück.

12. Übersetzung nach Howald-Meyer (Anm. 9) 53.

13. Zu den Wohnsitzen der Raeter – Strabo beschreibt hier die Wirtschaftsform dieses Volkes – s. R. Frei-Stolba, 'Die Räter in den antiken Quellen', in: Das Räterproblem in geschichtlicher, sprachlicher und archäologischer Sicht, *SRMC* 28 (1984) 6-21.

14. E. Meyer, 'Die geschichtlichen Nachrichten über die Räter und ihre Wohnsitze', *JbSGU* 55 (1970) 125.

15. Plin. *HN* 8.179: *(boues) non degeneres existimandi etiam minus laudato aspectu: plurimum lactis Alpinis, quibus minimum corporis, plurimum laboris capite, non ceruice iunctis.*

16. Plin. *HN* 11.240: *Laus caseo Romae...duobus Alpes generibus pabula sua adprobant: Delmaticae Docleatem mittunt, Ceutronicae Vatusicum*, vgl. dazu Howald-Meyer (Anm. 9) 77 mit Kommentar. S. auch die Anekdote S.H.A. *Ant. Pius* 12.4: *mors autem eius (sc. Antoninus Pius) talis fuisse narratur: cum Alpinum caseum in cena edidisset auidius, nocte reiectauit atque alia die febre commotus est.*

17. Zu den Käsesorten s.E. P. Herdi, *Die Herstellung und Verwendung von Käse im griechisch-römischen Altertum*, Diss. Bern (1918) 38 ff. Freilich darf nicht jeder Hinweis auf Käse im antiken Schrifttum mit Alpwirtschaft in Verbindung gesetzt werden, so aber etwas undeutlich Grass (Anm. 7) 241, da die Römer selbst ebenfalls Käse herstellten, dazu Herdi 24 ff., 27 ff. mit Hinweis auf Varro *RR* 2.11.3.

18. So Herdi (Anm. 17) 11 mit Anm. 7 nach einer humanistischen Konjektur bei H. Blümner, *Die römischen Privataltertümer* (1911) 583 Anm. 14, was in der moderneren Literatur teilweise übernommen wurde. S. zur Frage nun ausführlich G. E. Thüry, 'Antike Textzeugnisse über kleinwüchsige Rinder im Alpengebiet und im freien Germanien', in: H.-M. von Kaenel und M. Pfanner, *Tschugg – Römischer Gutshof, Grabung 1977* (1980) 102 ff. bes. Anm. 157. Anders etwa H. Krämer, *RE* Suppl. VII (1940) Sp. 1168 s.v. 'Rind'.

19. Columella, *RR* 6.24.5: (Fütterung der frischgeborenen Kälber) *et tener uitulus torrido molitoque milio et permixto lacte saliuatur. melius etiam in hos usus Altinae uaccae probantur, quas eius regionis incolae ceuas appellant...* Zum System der Ammenkühe ebenda. S. auch W. Richter, *L. Iunius Moderatus Columella, Zwölf Bücher über Landwirtschaft*, Slg. Tusculum (1982) Bd.II, 83 mit Anm. 15 zu Altinum, Landstadt in der Nähe des heutigen Venedig. Altinum war ebenfalls durch Schafzucht berühmt, s.J. M. Frayn, *Sheep-Rearing and the Wool Trade in Italy during the Roman period*, (1984) 37. Zu *ceua*, möglicherweise ein venetisches Wort, s.A. Walde- J.B. Hofmann, *Lateinisches etymologisches Wörterbuch*, (1938) 209. Thüry (Anm. 18) 102 Anm. 157 hält beide Interpretationen für möglich.

20. S. oben Anm. 16.

21. Plin. *HN* 17.69: *eadem causa et trimenstre (sc. triticum) inuenit detinentibus terras nivibus, quod tertio fere a satu mense, cum et in reliquo orbe, metitur. totis hoc Alpibus notum, et hiemalibus provinciis nullum hoc frumento laetius, unicalamum praeterea nec usquam capax, seriturque non nisi tenui terra.*,. s. auch Howald-Meyer (Anm. 9) 79, zweifellos als Getreide zur Selbstversorgung zu verstehen.

22. Plin. *HN* 16.190, 197, 200; Tiberius liess zur Wiederherstellung einer verbrannten Schiffkampfbrücke in Rom Lärchen aus Raetien kommen, deren Grösse sehr bewundert wurde. Die vollständige Zusammenstellung der literarischen Nachrichten bereits bei J. Heierli und W. Oechsli, 'Urgeschichte Graubündens mit Einschluss der Römerzeit', *Mitt. Antiquar. Gesell. Zürich* 26 (1903) 1-80, bes. 49.

23. Vermutlich wilder Honig oder dann aus Bienenzucht, die nebenher betrieben wurde, so auch Stauffer-Isenring (Anm. 3) 129. In der Wirtschaftsform der römischen *uilla* war Bienensucht anderseits ein Zweig der *pastio uillatica*, der intensiv betrieben wurde und eine hohe Rendite abwarf, s. W. Rinkewitz, *Pastio Villatica, Untersuchungen zur intensiven Hoftierhaltung in der römischen Landwirtschaft*, Europ. Hochschulschr. R. 3, Bd. 234 (1984) 119 f.

24. S. Van Berchem (Anm. 2) 209. Die Gesteine benötigten für grössere Transportvolumina die Wasserwege, so dass die Gebirgsgesteine eher nur lokal verwendet wurden, s. auch F. Braemer, 'Industrialisation et romanisation des régions de montagne', *Atti* (Anm. 2) 79-88. Zur Ausbeutung der Metallvorkommen, die hier nicht behandelt werden kann, s. bes. Pauli (Anm. 2) 271 ff.; das Hauptgebiet war Noricum.

25. Claud. *Bell. Poll.* 5.355: *pastorali...sub culmine*; 5.362 f.: *Alpinae...casae*; 5.349 f.: *nulla Lyaei/ pocula, rara Ceres*; Uebersetzung bei Howald-Meyer (Anm. 9) 153 f.

26. S. dazu allg. I. Chirassi Colombo, 'Acculturazione e morfologia di culti alpini', *Atti* (Anm. 2) 160 mit Anm. 6.

27. Nach G. Walser könnte die bekannte Inschrift CIL XII 113 über die Grenzziehung zwischen den Ceutrones und den Viennenses, die Cn. Pinarius Clemens vornahm, die Schlichtung eines Weidestreites bedeutet haben, s. dens., *Via per Alpes Graias, Beiträge zur Geschichte des kleinen St. Bernhard-Passes in römischer Zeit*, Historia Einzelschr. 48 (1986) 80.

28. Ausser Graubünden und dem hier nicht untersuchten Wallis haben die übrigen alpinen Gebiete der Schweiz keine epigraphischen Zeugnisse geliefert. Immerhin ist in diesem Zusammenhang noch an den Tempelbezirk von Thun/Allmendingen zu erinnern, der die singuläre Weihinschrift an die *Alpes*, die Alpengottheiten, bewahrt hat, s. Howald-Meyer (Anm. 9) Nr. 234: *Alpibus/ex stipe/reg(io) Lind(ensis)*; zu den weiteren Funden und Ausgrabungen s. den Forschungsstand bis 1964 bei H. Bögli, in : Chr. Strahm u.a., *Ur-und Frühgeschichte der Gemeinde Thun* (1964) 52 ff.; zu den späteren Ausgrabungen H. Grütter, 'Einzigartige Weihegaben aus dem gallorömischen Tempelbezirk von Thun-Allmendingen', *HA* 7 (1976) 102 ff. Wie auf den anderen Pässen sind auch auf dem Julier Reste eines Heiligtums gefunden worden, s. die Forschungsbersicht mit Lit. bei Overbeck (Anm. 3) 133-135. R. Wyss, 'Die archäologische Erforschung des Oberhalbsteins', *HA* 8 (1977) 5 möchte das Inschriftenfragment *A[---]/RE[---]* nach dem Stein von Allmendingen ergänzen, dagegen jedoch mit Recht F. E. Koenig, 'Der Julierpass in römischer Zeit', *JbSGU* 62 (1979) 83.

29. S.A. Bruckner, H. Erb, E. Meyer, 'Römische Votivaltäre aus dem Engadin und neue Inschriften aus Chur', *Helvetia Antiqua, Festschrift E. Vogt* (1966) 223-232.

30. Plin. *HN* 36.159. Die Altäre sind etwa 40-47 cm hoch und einfach verziert. Zu den Weihungen an *Mercurius Cissonius* s. unten Anm. 37.

31. T. Tomăsević-Buck, *JbSGU* 68, 1985, 242-244: *I(oui) O(ptimo) M(aximo)/Marti(us)/ Magian(us)/ex uoto* mit den verschiedenen Lesungsvarianten von H. Lieb.

32. Die Altäre wurden beim Aushub der Baugrube für ein Ferienhaus am Nordwestrand von Sils-Baselgia gefunden, wobei die Untersuchung der Schicht, in der die Altäre lagen, ergab, dass der Silsersee offenbar in römischer Zeit grösser gewesen war, s.H. Erb (Anm. 29) 223.

33. Zur Namensform Meyer (Anm. 29) 227 f. *Valerius* ist ein in Oberitalien sehr geläufiger, oft auch als Individualname gebrauchter Name, s.J. Untermann, 'Namenlandschaften in Oberitalien', *BN* 12 (1961) 27, dort auch Belege zu dieser peregrinen Namensform; zum im keltischen Sprachbereich häufigen Namen *Tertius* s. G. Alföldy, *Die Personennamen in der römischen Provinz Dalmatia* (1969) 307. Gerold Walser sieht im Stifter freilich eher einen Freigelassenen, der das Transportwesen längs der römischen Hauptstrassen, hier der Route von Chiavenna-Julier-Chur-Bodensee, besorgte und sich mit seinen vier Weihinschriften an die hier hauptsählich verehrten Götter wandte, womit den Zeugnissen ein stärkerer römischer Charakter zukäme, G. Walser, *Römische Inschriften in der Schweiz* (1979/80) Bd. III Nr. 309-213, ähnlich die Lesung des Votivsteines an *Mercurius Cissonius*, unten Anm. 37, womit in dieser Gebirgsgegend auffallend viele *liberti* bezeugt wären.

34. D. van Berchem, 'Du portage au péage. Le rôle des cols transalpins dans l'histoire du Valais celtique', *MH* 13 (1956) 199 ff.; jetzt in: *Les routes et l'histoire* (1982) 67-78.

35. CIL XIII 5243 (= Howald-Meyer, Anm. 9, Nr. 261; Walser, Anm. 33, II Nr. 192). Kleine Kalksteinplatte mit schlechter Schrift, gef. 1868: *Deae Dianae/et Siluano/ursari/posueru/nt ex uoto.* Übrigens kommt die Zusammenstellung der drei Götter Diana, Merkur und Silvanus auch einmal auf einer Inschrift in Dalmatien vor, so bereits Meyer (Anm. 29) 228, übernommen von Pflaum im Kommentar AE 1966, 275. Die dalmatinische Inschrift AE 1933, 73 ist recht zerstört, aber mit einer Reliefdarstellung der drei Gottheiten geschmückt.

36. CIL IX 3375. Meyer (Anm. 29) 228 mit Anm. 23, sehr zurückhaltend; Pflaum, Kommentar zu AE 1966, 275.

37. Howald-Meyer (Anm. 9) Nr. 28 = Walser (Anm. 33) III Nr. 306: *Mercurio/Cissonio/pro bon[o]// comm[uni]//[---]*, verbesserte Lesung nach H. Lieb, *Lexicon Topographicum der römischen und frühmittelalterlichen Schweiz* (1967) 102; die beiden früheren Herausgeber lasen nach R. Laur-Belart, JbSGU 31 (1939) 90: *pro bon[o]/Çamį[lli]*, was zu kühnen gedanklichen Verbindung mitr den Camillern von Aventicum reizen konnte. Lieb setzt eine geläufige Formel ein. Walser (Anm. 33) III Nr. 307: *Mercurio/[Ci]ssonio/[M]a[t]utino/Valerius/[Ge]rmani-* Walser ergänzt auch hier *[l(ibertus)]*, s. auch Meyer (Anm. 29) S. 231 Anm. 19 (Lit.). Die Weihaltärchen (16 cm und 11 cm hoch) wurden 1938 und 1958 bei der Talsperre bei Promontogno (Murus) in neuerem Mauerschutt in sekundärer Verwendung gefunden.

38. *Mercurius Cissonius* darf dabei sicherlich als 'Wagengott', als Gott der Fuhrleute, interpretiert werden, der übrigens ebenfalls in Aventicum und in den Rheinlanden verehrt wurde, so bereits Meyer (Anm. 9) Nr. 28 Kommentar und Walser (Anm. 33) III Nr. 306. Die Inschrift aus Aventicum: CIL XIII 11 476. Begründeterweise wird im Beinamen *Cissonius* eine Anspielung auf das gallische Wort *cisium*, auch *cissum* (zweirädriger Wagen) gesehen, womit der Zusammenhang mit den Fuhrdiensten offenkundig wird, der sich auch in den lokalen Ortsnamen wiederfindet. S. zu Tarvessedum Lieb (Anm. 38) 132-134, auch Meyer (Anm. 9) Nr. 28 Kommentar (eig. 'Ochsenwagen'); zu Tinnetio Lieb, 139-142, Meyer a.a.O. (eig. 'Anspannung' der Wagen mit Ochsen) nach älterer sprachwissenschaftlicher Literatur.

39. Overbeck (Anm. 3) 195 f.: 'Von einer Rekonstruktion des mittelkaiserzeitlichen Siedlungsbildes zwischen Bregenz und den Pässen im Süden sind wir noch weit entfernt', anschliessend der Hinweis, dass sich seit den Grabungen von J. Rageth auf dem Padnal bei Savognin (Regelmässige Ausgrabungen seit 1971, s. die Berichte seit *JbSGU* 59 (1976) 123 ff.) die Funde in der Umgebung häufen, bzw. Funde eben gemeldet werden; ähnlich Wiblé (Anm. 1) 190: 'découvertes fortuites qui ont été signalées'.

40. S. oben Anm. 3. Es handelt sich hier um einen Gutshof, möglicherweise eine *mutatio*.

41. Dazu s. jetzt die umfassende Bibliographie *IVS Inventar historischer Verkehrswege der Schweiz*, hsg. v.K. Aerni und H.E. Herzig, 2 Bde (1982) 2. erw. Aufl. 1984; s. auch die Uebersicht von H.E. Herzig, 'Die Erschliessung der Schweiz durch die Römer', *Historische und aktuelle Verkehrsgeographie der Schweiz*, hsg. v.K. Aerni und H.E. Herzig (1986) 5-21.

42. Immerhin sind Fortschritte zu verzeichnen; s. etwa für das Engadin Nauli bei Conrad (Anm. 3) 61 zu Zernez, wo anlässlich einer Friedhoferweiterung Ueberreste eines römischen Gebäudes festgestellt wurden. Nach Zeitungsberichten wurden ebenfalls Funde am Iffigensee (Simmental, BE) getätigt. Es würde sich lohnen, nur schon die bisher gemachten römerzeitlichen Funde nochmals genau zu kartieren.

43. R. Wyss, 'Die Eroberung der Alpen durch den Bronzezeitmenschen', *ZAK* 28 (1971) 130-45; ders., 'Die Höhenfunde aus dem Fürstentum Liechtenstein', *HA* 9 (1978) 137-44. Pittioni (Anm. 7) 181-183 mit Hinweisen auf ältere Spezialaufsätze; dazu M.R. Sauter, 'L'occupation des Alpes par les populations préhistoriques', in: *Histoire* (Anm. 2) 61-94, bes. 76 f. weitere Literatur bei Grass (Anm. 7), bes. 230 ff.

44. Zu diesen beiden Voraussetzungen s. Grass (Anm. 7) 230. Da hier Alpwirtschaft und Viehzucht im Vordergrund stehen, kann der zweite Problemkreis, die Suche nach den Metallen und die daraus folgende Erkundung der höher gelegenen Gebiet weggelassen werden, zu diesen Zusammenhängen s. Pauli (Anm. 2) 271 ff. und Grass (Anm. 7) 235 ff. zu den Bergbausiedlungen in den Ostalpen. Wie Van Berchem (Anm. 2) 209 zu Recht bemerkt, spielten diese Vorkommen, ausser in Noricum, in der römischen Zeit eine geringere Rolle.

45. Die einzelnen Funde kartiert bei Wyss 1971 (Anm. 43) 133, 135, wo er versucht, die verschiedenen Kategorien der Funde, die Höhenstationen im Berner Oberland (z.B. die Funde aus dem Simmental im Mamilchloch, Schnurenloch usw.), die neben Bronzen auch Keramikreste und Tierknochen enthielten und damit auf eine periodische Besiedlung schliessen lassen, von den isolierten Weihegaben zu trennen. Auch der berühmte Fund von Erstfeld wird als Opfergabe interpretiert, s. R. Wyss, *Der Schatzfund von Erstfeld. Frühkeltischer Goldschmuck aus den Zentralalpen* (1975).

46. J. Hubschmid, *Die Alpenwörter romanischen und vorromanischen Ursprungs* (1951); darnach Hinweise bei Grass (Anm. 7) 244 ff. mit Überblick über die Forschungsgeschichte; auch König-Ploss (Anm. 7) 184-185. Offenbar müssen verschiedene Sprachschichten mit ganz frühen, keltischen und lateinischen Wörtern voneinander getrennt werden. Zur ganz frühen Schicht J. Hubschmid, *Mediterrane Substrate*, Romanica Helvetica Ed. 70 (1960) bes. 80 ff.

47. So Wyss 1978 (Anm. 43) 143; zum Klimasturz auch P. Guichonnet, in: *Histoire* (Anm. 2) 29 und Pauli (Anm. 2) 270.

48. S. Lamb 1977 (Anm. 5) 424 f. mit Hinweis auf Saserna (Vater oder Sohn) bei Colum. I, 1, 4, der die Klimaverbesserung im Zusammenhang mit Oliven und Weinpflanzungen erwähnt; dazu G. Delibrias, M. Le Roy Ladurie und E. Le Roy Ladurie, 'La forêt fossile de Grindelwald: Nouvelles datations', *Annales ESC* (1975) 137-47, die die naturwissenschaftlichen Erkenntnisse historisch umsetzen.

49. S. Birkenhauer (Anm. 7) 103; der Ausbau der Alpwirtschaft durch Rodung erfolgte erst im späten Mittelalter.

50. So Grass (Anm. 7) 247 ff.; zu den Heidenhüttli in der Forschung ausführlich J. Hösli, in: *Bergeten* (s. nächste Anm.) 55 f.

51. W. Geiser (Hsg.), *Bergeten ob Braunwald, ein archäologischer Beitrag zur Geschichte des alpinen Hirtentums* mit Beiträgen von M.-L. Boscardin, M. Gschwend, J. Hösli, S. Meyer und W. Meyer. Zu

den Wüstungen s. bes. W. Meyer, 'Wüstungen als Zeugen des mittelalterlichen Alpwesens', *SZG* 29 (1979) 256-264 (skeptisch gegenüber der prähistorischen Alpwirtschaft); M. Gschwend, 'Alpine Wüstungen', *Festschrift R. Wildhaber* (1973) 212-215.

52. H. Boesch, 'Nomadismus, Transhumanz und Alpwirtschaft', *Die Alpen* 27 (1951) H. 6, 202-207; S. Berezowski, 'Typologie des migrations pastorales en Europe et méthodes de leurs études', in: *L'aménagement de la montagne, compte-rendu du IIIe colloque franco-polonais de géographie, mai 1969* (1971) 165-74. S. auch W. Gallusser und H. Haefner, 'Eine internationale Arbeitsgruppe für die geographische Terminologie der Agrarlandschaft', *Geographica Helvetica* 20 (1965) 228-29, wo eine einheitliche Terminologie gefordert wird. Zur Alpwirtschaft und Transhumanz s. auch Zöbl (Anm. 8) bes. 2 ff., und R. Gubler-Gross, *Moderne Transhumanz in der Schweiz*, Diss. Zürich (1962) bes. 2 f.

53. Boesch (Anm. 52) 203.

54. Boesch (Anm. 52) 204; Birkenhauer (Anm. 7) 106 mit weiteren Differenzierungen; zur Alpwirtschaft liegt eine reiche Forschung vor, die sich, abgesehen von den modernen Anleitungen zum Alpwesen, vor allem mit einzelnen Gebieten beschäftigt, s. bes. R. Weiss, *Das Alpwesen Graubündens* (1941).

55. Stauffer-Isenring (Anm. 3) 128 führt einige mögliche Spuren an, auch wenn sie zweifellos zu Recht bemerkt, dass die Waldweiden bei der Ernährung des Viehs (Rinder und Schafe sowie Ziegen) im Frühling und Herbst eine grosse Rolle spielten.

56. Zöbl (Anm. 8) 32 nach H. Schmid, *Die Oberengadiner Land-und Alpwirtschaft*, Diss. Zürich (1955); Gubler-Gross (Anm. 54) 11.

57. Gubler-Goss (Anm. 52) 6, 16 ff.

58. Zu den Wirtschaftsformen der *uilla* s. die gute Zusammenfassung von R. Fellmann, 'Die römische villa rustica ökonomisch gesehen', in: *Technik und Wirtschaft in ur- und frühgeschichtlicher Zeit*, Einführungskurse in die ur-und frühgeschichtliche Archäologie der Schweiz (1983) 75-80, auch frz. in: *Economie et techniques de la préhistoire aux origines du Moyen-Age* (1984). Dazu W. Drack, 'Die Gutshöfe', *UFAS* V (1975) 49-72, bes. 68-70 zu den Wirtschaftsgebäuden.

59. S. Fellmann (Anm. 58) 80 mit Abbildungen unpublizierten Materials aus der Villa Seeb (Glocken und Kastrierzange), zu dieser Villa s.W. Drack, Der römische Gutshof bei Seeb, Archäologische Führer der Schweitz 1 (1969); auch R. Fellmann, in: A.G. McKay, *Römische Häuser, Villen und Paläste*, dt. Ausgabe bearbeitet und erweitert von R.F. (1980) 193-205; aus der Villa Seeb sind nach Drack, UFAS V (1975) 70 zwei Käsemodel erhalten. Vgl. auch L. Flutsch, *La villa gallo-romaine d'Orbe-Boscéaz, état des questions et directions de recherche*, Mém. de licence, Université de Lausanne (1986) 104 (Glocke; Käsekorb).

60. S. die Zusammenstellungen von H.-R. Stampfli, 'Die Tierreste aus der römischen Villa von Ersingen-Murrain in Gegenüberstellung zu anderen zeitgleichen Funden aus der Schweiz und dem Auslande', *JBM* 45/46 (1965/66) 449-69; dazu die weiteren Analysen der Tierknochenfunde von E. Schmid und M. Hummler bei: S. Martin-Kilcher, *Die Funde aus dem römischen Gutshof von Laufen-Müschhag*, (1980) 80ff. H.R. Stampfli bei: H.-M. von Kaenel und M. Pfanner, Tschugg-Römischer Gutshof, Grabung 1977 (1980) 97-110.

61. Drack, *UFAS* V (Anm. 61) 58, 70.

62. E. Schmid, 'Tierreste aus einer Grossküche von Augusta Raurica', *Basler Stadtbuch* (1967) 176-86; zu den Ratschlägen der römischen Agrarschriftsteller s. H. Dohr, *Die italischen Gutshöfe nach den Schriften Catos und Varros*, Diss. Köln (1965) 66 ff.

63. R. Laur-Belart, 'Gallische Schinken und Würste', *Ur-Schweiz* 17 (1953) 33 ff.; U. Müller, 'Die römischen Gebäude in Kaiseraugst-Schmidmatt', *ArchS* 8 (1985) 19; A.R. Furger, 'Vom Essen und Trinken im römischen Augst', *ArchS* 8 (1985) 168-87.

64. S. vor allem K.-H. Knörzer, 'Römerzeitliche Pflanzenfunde aus Neuss', *Novaesium IV* (1970) bes. 146-9, ebenda 148 auch die Annahme eines Anbauwechsels in zweijährigem Rythmus, wobei sich Fruchtbau und Brache mit Weidegang abwechselten. Diese Fragen verdienten eine neue eingehende Untersuchung der einschlägigen antiken Literatur. S.U. Körber-Grohne, *Nutzpflanzen und Umwelt im römischen Germanien*, Kl. Schr. z. Kenntnis d. röm. Besetzungsgeschichte Südwestdeutschl. (1979) 62; S. Fünfschilling, Chr. Jacquat, J. Schibler und A. Zürcher, 'Pflanzenanbau, Nahrungsmittel und Essgewohnheiten im römischen vicus Vitudurum-Oberwinterthur', *ArchS* 8 (1985) 160-167, bes. 164 f.

65. N. Morard, 'L'élevage dans les préalpes fribourgeoises: des ovins aux bovins (1350-1550)', in: *Actes du colloque intern.: L'élevage et la vie pastorale dans les montagnes de l'Europe au moyen âge et à l'époque moderne* (1984) 15-26. S. auch A. Hauser, *Wald und Feld in der alten Schweiz* (1972) 7-21.

66. J. Harmand, 'Le groupe des Viehwege vosgiens: une zône d'élevage gallo-romain en basse montagne', *Actes du colloque* (Anm. 65) 203-15.

67. S. bes. die Forschungsberichte und Uebersichten über die Funde im Kanton Freiburg von H. Schwab, Chronique archéologique 1980-1982, Editions universitaires (1984) sowie *Histoire du canton de Fribourg*, auch dt.: *Geschichte des Kantons Freiburg* Bd. I (1981) mit den Beiträgen von H. Schwab und T. Zawadzki.

68. Für das Schweizerische Mittelland ist bisher erst der noch nicht vollständig ausgegrabene Gutshof von Orbe-Boscéaz als grosse Besitzung zu nennen, s. vorläufig Flutsch (Anm. 59).

69. So weist W. Eck, 'Senatoren aus Germanien, Raetien, Noricum?' in: *Epigrafia e ordine senatorio* Bd. II, Tituli 5 (1982) 539-52 darauf hin, dass die gesamte Germania Superior nach den bisherigen Inschriftenfunden keinen Senator hervorgebracht hat.

LE PASTORALISME EN GAULE

Ce texte ne saurait prétendre à constituer un 'rapport de synthèse', tant est squelettique la documentation disponible en France (la Suisse faisant l'objet d'une communication propre). Si le thème de ce congrès avait concerné l'élevage, les donnés auraient été plus nombreuses, mais, à s'en tenir au strict pastoralisme, indissociablement lié à la transhumance et s'appliquant pour l'essentiel aux ovins, on se trouve dans l'embarras et ce d'autant plus qu'il est souvent impossible de distinguer élevage fixe et pastoralisme transhumant. Mais si la récolte doit être maigre, le sujet ne manque pas de vertus 'heuristiques' capables d'ouvrir des pistes de recherches fécondes.[1]

1. *Sources écrites et figurées*

Stricto sensu, deux textes (deux seuls) peuvent évoquer le pastoralisme. Le premier est ambigu et ne concerne que marginalement la Gaule: Strabon 4. 6. 2, dit des Ligures qu''ils vivent de leurs troupeaux, de lait et d'une boisson à base d'orge; ils occupent les terrains du bord de mer et surtout les montagnes'. Le second est plus explicite: Pline, *N.H.* 21. 57, évoque le cas de la Crau: 'les Plaines-de-pierre, dans la province Narbonnaise, sont aujourd'hui remplies de thym; c'est presque leur seul revenu, des milliers de moutons y venant de régions lointaines paître ce thym', ce que Strabon 4. 1. 6, confirme en termes plus vagues, sans allusion à la transhumance: la Crau 'est recouverte de pierres grosses comme le poing sous lesquelles pousse un chiendent qui fournit une abondante pâture au bétail.' Notons que le texte de Pline demeure muet quant à la provenance des troupeaux: les Alpes, les Préalpes, les Cévennes? On ne sait.

Toutes les autres citations[2] font référence à l'élevage sans guère de précision et reprennent le leitmotiv selon lequel la Gaule est une terre de pâture (*pecuaria*: Sid. Apoll. *Carm.* 5. 45; *terra est frumenti praecipue ac pabuli ferax*: Mela 3. 17). Pour les ovins, les vêtements de laine attestés par Strabon (d'après Posidonius) pour la protohistoire (4. 4. 3) devinrent ensuite célèbres et recherchés (Martial 6. 11. 7: *me pinguis Gallia vestit*) et les références sont nombreuses,[3] de même que ne manquent pas dans les tomes XII et XIII du *CIL* les inscriptions relatives aux métiers de la laine, de sa transformation à sa commercialisation.[4] Mais jamais ne sont évoqués de pasteurs ou de bergers à une exception près-d'ailleurs germanique-liée à un fait-divers: un *pecuarius* assassiné par son esclave.[5] En outre, contrairement à une tradition bien ancrée dans certaines histoires régionales, *aucun* fromage de brebis ou de chèvre n'est attesté avec certitude.[6]

La recension des représentations d'animaux liés à la chasse ou à l'élevage en Gaule (petites espèces exceptées) peut être faite d'après le *Recueil général des bas-reliefs, statues et bustes de la Gaule romaine* d'E. Espérandieu. Sur près de 400 pièces, les pourcentages s'établissent ainsi:

cheval	28%
taureau (ou tête de taureau)	22%
bélier (ou tête de bélier)	14%
sanglier	12,5%
chèvre	5%
bouc	4,5%
boeuf (ou bucrâne)	4,5%
brebis, mouton	2%
porc	2%

à quoi on ajoutera, parmi les 'divers', 3% correspondant à la représentation de bergers. La part des ovins et des pasteurs est d'autant plus négligeable qu'elle répond le plus souvent à une mise en scène 'alexandrine': le berger, sa flûte, ses brebis, sa cabane.[7] Mais il est vrai que la représentation de chevaux (liée à des militaires ou à Epona), de protomés divers, de scènes de sacrifice, etc, répondent également à des schémas conventionnels.

A s'en tenir à cette catégorie de sources, les conclusions sont vite tirées: il existait en Gaule un élevage de moutons fournissant une abondante matière première pour la fabrication de vêtements; une transhumance est attestée pour la Crau. Et l'on ne sait rien de plus.

2. *Sources archéologiques*

Sur de nombreux sites protohistoriques ou gallo-romains, on a fait moisson d'objets en relation avec les ovins: forces pour tondre les moutons, faisselles en céramique pour la fabrication des fromages, râcloirs, poinçons et aiguilles de bronze pour préparer les peaux et les assembler, fusaïoles et pesons pour le traitement de la laine, traces de la transformation des os et des cornes, etc. Mais, outre que beaucoup d'instruments ont disparu (par exemple, ceux en bois et en vannerie, qui jouent un rôle important pour la traite et le fromage), on ne dispose pas de méthode fiable qui permettrait une estimation (ou, au moins, une approximation) quantitative.

En dépit des progrès de la palynologie, c'est l'étude des ossements animaux (trouvés en fouille) qui constitue le principal apport des dernières années -du moins dans certains pays et pour l'archéologie des périodes 'classiques'. Certes, la méthodologie ne se met pas au point facilement[8] et la validité de l'interprétation repose sur de nombreux facteurs,[9] si bien que la prudence invite à tirer de ces analyses non des enseignements 'absolus' mais des comparaisons: à conditions de fouilles à peu près semblables, dans des milieux géographiques et culturels à peu près cohérents, l'étude des rebuts offre une idée approximative des espèces consommées, du moins pour les plus importantes à la fois en taille et en quantité.

Nous ne disposons malheureusement que de très rares monographies ou articles de synthèse.[10] En 1976, Th. Poulain avait défini, pour la France, quinze 'groupes régionaux' correspondant à des régions naturelles; à l'intérieur de chacun d'eux, d'après le comptage du nombre d'ossements qu'on lui avait donné à expertiser, et en calculant le nombre minimum d'individus que ces os représentaient, elle proposait une 'hiérarchie' entre les espèces consommées: telle région consommait plus de boeuf que de mouton ou de porc, etc.[11] Ce mode de calcul avantage les petites espèces; c'est pourquoi, par un transfert un peu arbitraire, on a tendance aujourd'hui à 'transformer' ce nombre minimal d'individus en 'poids de viande consommable': selon son âge, le mouton 'pésera' de 6 à 17 kg, le porc d'environ 13 à 48 kg et le bovin de 52 (pour un veau d'un mois et demi) à 180 kg (pour un adulte moyen). Ne nous étendons pas sur ces calculs[12] qui ne nous intéressent que pour cadrer un débat précis.

Selon Th. Poulain, une seule région française avait connu, durant tout le cours de son histoire, une prédominance constante des ovi-capridés (moutons et chèvres -et leurs conjoints): celle que composent le Languedoc et la Provence.[13] Ces dernières années, Ph. Columeau, étudiant des dizaines de milliers d'os trouvés dans des gisements archéologiques stratifiés et bien datés, confirme et infirme ces conclusions: s'il est vrai qu'en *nombre* de restes les ovi-capridés tiennent toujours le premier rang (ou presque, l'exception s'expliquant par la nature du gisement), en revanche, lorsqu'on calcule en 'poids de viande', ils se trouvent souvent égalés ou dépassés par les 'suidés' et toujours surpassés par les bovidés (de loin: en moyenne, près de trois fois). Or, ce n'est pas un phénomène conjoncturel: le répartition des espèces connaît (à quelques modulations près qui peuvent provenir de l'échantillonnage) une remarquable stabilité du néolithique (en tout cas, de l'Âge du Bronze) à la fin de l'Antiquité.[14] Voilà qui suscite la réflexion.

3. *Les traditions*

Dans la vie (qui inclut l'existence scientifique), il faut faire des paris et prendre des risques: acceptons donc les résultats de ces analyses paléo-zoologiques que nous venons de prendre à notre compte. Mesurons-en le scandale: la Provence et le Languedoc, qu'on associe toujours aux moutons et aux chèvres (impossible, dans une crèche provençale, de trouver d'autres animaux que les ovins, le chien et l'âne ou le mulet – le boeuf ne tenant qu'un rôle: se tenir près de l'enfant Jésus), seraient en fait des pays où les bovidés auraient tenu une place éminente dans l'alimentation?

Certes, Posidonius (si Strabon l'a suivi) n'a vu dans le Midi ni vaches ni boeufs, mais Strabon n'en signale pas davantage dans le reste de la Gaule – ou du moins il ne juge pas utile d'en faire état, se contenant d'une formule générale (4. 1. 2: 'en Gaule prospèrent toutes les espèces de bétail d'élevage'). En revanche, il met l'accent sur l'étrangeté des porcs semi-sauvages (*ibid.*). On sait bien, par ailleurs, qu'un troupeau de moutons en déplacement attire plus le regard qu'une quinzaine de vaches au pré, même si les uns et les autres représentent pour les paléo-zoologues le même 'poids de viande'. Inutile de nous appesantir sur une ethnographie 'pittoresque' dont on aurait tant d'exemples en Gaule même.[15]

Le vrai problème est ailleurs: l'exemple méridional nous amène à nous interroger sur la validité d'un processus classique (qui ne concerne pas que la Gaule). Dans la plupart des livres ou contributions consacrés au pastoralisme médiéval ou moderne, on voit en lui la continuation (ou la reprise) d'un phénomène antique et développé à peu près à la même échelle. Comment trancher? Il faudrait évidemment pouvoir comparer des sources ... comparables.[16] Faute de mieux, on ne peut confronter aux analyses paléo-zoologiques portant sur l'Antiquité que les études qui se sont consacrées à l'alimentation des époques médiévale et moderne, et elles ne sont pas bien nombreuses.[17]

Dans le Midi, il semble que la consommation de viande de mouton ait été relativement plus importante que dans l'Antiquité. Aux XIVe et XVe siècles, elle se trouve à peu près à égalité avec la consommation de boeuf (et bien au-dessus du porc) et sa part ne cesse de grandir: au milieu du XIXe, elle est prééminente. On est tenté de partager les vues d'A.-M. Brisebarre plaçant l'essor de la transhumance méridionale au XIIIe siècle et son plein développement au XVe, l'effondrement correspondant à l'extension de la culture de la vigne, et surtout à l'épidémie de phylloxéra qui amena l'introduction de la vigne américaine.[18]

Autrement dit: cet exemple méridional nous conduit à penser que, faute de preuves, nous utilisons sans trop de scrupules l'argument 'régressif' que nos collègues médiévistes et modernistes nous offrent. Nous prenons aisément à notre compte ce pastoralisme qu'on nous remet en cadeau, d'autant plus volontiers que la Grèce ou l'Italie nous offrent des alibis faciles. Pourtant, des réticences se font jour en diverses régions de France.[19] Pour une raison simple: les enjeux du pastoralisme sont moins indifférents qu'il n'a longtemps paru.

4. *Les enjeux et les perspectives*

Si l'on s'en tenait au simple phénomène du pastoralisme transhumant, nos erreurs d'appréciation ne seraient pas plus graves que toutes celles que nous commettons sur l'architecture, l'épigraphie, la céramique, etc. Mais ce sujet a une particularité: il porte avec lui toutes sortes de connotations qui le dépassent: idéologiques, sentimentales, religieuses, etc, qu'on ne cerne que malaisément et qui, curieusement, lorsqu'on tente de les analyser, nous détournent du pastoralisme lui-même.[20]

Pour la Gaule, les enjeux sont aussi d'ordre historique: le pastoralisme (entendu au sens large de mouvements de pasteurs, d'éleveurs) est souvent présenté comme le moteur de faits de civilisation essentiels. Pour certains, c'est lui qui aurait poussé, vers le XIVe siècle avant J.-C., des 'proto-Celtes' d'Allemagne du Sud vers le Bassin Parisien et le Sud-Ouest de la France.[21] Pour d'autres, il aurait joué un rôle essentiel dans les 'mouvements celtiques' des environs de 500.[22] Des chercheurs méridionaux le placent à l'origine de courants commerciaux et d''acculturation' entre les rivages méditerranéens, les Cévennes et le Massif Central durant la protohistoire.[23] D'autres lui attribuent le tracé de certaines voies romaines qui auraient repris le trajet de 'drailles' de transhumance.[24]

Voilà qui nous conduit au coeur du problème: le pastoralisme ne constitue-t-il pas souvent une réponse (un alibi?) à toutes sortes de questions dérivées d'une absence de données? Ironisons un peu: vous trouvez des nécropoles sans habitats qui leur correspondent? Vous avez repéré des tumulus isolés, des gisements de hauteur, vous n'avez pas d'explication? Une solution: le pastoralisme![25] Solution d'autant plus aisément adoptée que ne manquent pas en France certaines formes d'habitats semblables aux 'bories' méridionales ou aux 'cadoles' bourguignonnes dont le principal mérite est d'autoriser des datations s'étalant du néolithique à... hier.

L'ironie ne suffit pas. On peut déployer les objections que s'attire tout argument *ex silentio*. Des nécropoles isolées? Lorsque J. Déchelette écrivait son *Manuel*, au début du siècle, il ne connaissait aucun habitat 'celtique' à part quelques *oppida* 'césariens'; aujourd'hui nous avons retrouvé des villages, des maisons, des agglomérations se dessinent et ce que Déchelette a écrit sur l'habitat doit être réécrit. Donc, lorsqu'on explique la présence de tel ou tel gisement peu caractérisé par une 'circulation' de pasteurs, la méfiance doit être de mise. D'autant qu'on a fort mal exploré en France les 'hauts pays': comment prospecter ou fouiller là où les vestiges sont rendus plus fragiles que les plus fragiles par les conditions climatiques?

L'apport des recherches préhistoriques constitue également un bon test pour l'argument *ex silentio*. En quelques années ont été accomplis des progrès saisissants. On sait désormais que les ovi-capridés existent en France depuis le mésolithique. Dès les débuts du néolithique, leur importance est très grande dans toute la France méditerranéenne. A cette époque, des grottes et des abris sous roche étaient utilisés comme bergeries, constituant des haltes pour le pastoralisme. Un pastoralisme à quelle distance? Les préhistoriens penchent vers de 'faibles' déplacements, liés à des 'niches écologiques' défavorables à l'agriculture (ou aux techniques agricoles peu efficaces des premiers paysans). Ces déplacements de transhumants se sont-ils accrus et quand? On ne sait, ou plutôt on trouve des problèmes semblables à ceux que nous avons signalés, ci-dessus: les sépultures dolméniques des Causses (le plus important noyau de France), les tumulus pyrénéens correspondent-ils à des ensevelissements de pasteurs? Les thèses prédominantes penchent plutôt en faveur d'une mobilité relativement faible et d'une spécialisation des 'secteurs'.[26]

C'est donc aux historiens de l'Antiquité de faire la preuve que des transformations se sont produites à l'Âge de Fer et à l'époque gallo-romaine. Or, on l'a dit, les analyses paléozoologiques concluent généralement à une stabilité de la répartition des espèces dans les secteurs où le pastoralisme a joué un rôle.[27] La multiplication (souhaitable) de ces analyses modifiera-t-elle le tableau? Peut-on compter sur la palynologie pour mettre en évidence, dès l'Antiquité, une déforestation des sommets pour créer des pâtures, phénomène qui n'est guère attesté pour l'instant qu'à l'époque médiévale (tardive) et moderne? Voilà qui suppose des campagnes systématiques accompagnées, si possible, de prospections et de fouilles. Vaste programme, mais ce n'est qu'à cette condition que nous pourrons opposer au silence de nos sources des données suffisamment fortes pour nous permettre de tirer des conclusions économiques et... d'échafauder éventuellement des théories historiques. Sans vouloir donner dans l'hyper-scepticisme, il convient de prendre conscience que la route risque d'être longue.

D'autre part, est-il sûr que la bonne méthode consiste à considérer le pastoralisme comme un phénomène en soi? Nos collègues médiévistes et modernistes ont montré les problèmes complexes que posent la complémentarité ou l'antagonisme non pas seulement (parfois) entre des populations (pasteurs, agriculteurs) mais entre des données économiques: par exemple, le fait que les moutons, dans les plaines, sont avant tout appréciés pour leur fumier mais qu'il faut que les troupeaux s'en aillent lorsque vient la saison des récoltes; l'emprise des défrichements (liés à l'extension, dans le Midi, des céréales, de l'olivier et de la vigne) se fait au détriment du pastoralisme (ce qui invite à la réflexion pour la période gallo-romaine); le rôle joué par le sel, qui est indispensable à la santé des moutons et à la fabrication des fromages mais qui coûte cher et qu'on ne saurait se procurer partout (ce qui n'est pas sans intérêt pour la protohistoire). La règle la plus importante peut être ainsi définie: 'les zones entre lesquelles les transhumants se déplacent ne sont pas uniquement vouées à l'économie pastorale: parfois l'une des deux, souvent toutes deux pratiquent en même temps l'économie agricole. Or, cette combinaison, qui souvent impose les migrations du troupeau, empêche toujours celles de la tribu. Le bétail va chercher ailleurs une partie de sa subsistance; les hommes sont retenus par les travaux des champs autour des maisons qui les fixent au sol'.[28] Des études médiévales, ressort aussi le rôle important joué par les monastères dans le développement du pastoralisme: comment apprécier l'ampleur du phénomène dans l'Antiquité tant que nous n'aurons pas trouvé de 'relais' comparables? En fait, c'est à une évaluation globale de l'occupation et de l'exploitation du sol qu'il faut demander la part du pastoralisme.

COLLÈGE DE FRANCE CHRISTIAN GOUDINEAU

NOTES

1. Pour la bibliographie générale, je ne citerai pas de nouveau *in extenso* les titres précédemment répertoriés dans l'introduction de P. Garnsey et qui concernent de près ou de loin la Gaule: Arbos 1922, Baticle 1974, Houston 1984, Lamboglia 1921; il est clair que la réflexion s'est beaucoup nourrie des contributions de M. Pasquinucci (1979 et 1984). Cités par J.F. Cherry et importants pour la méthodologie – ou pour la Gaule: Barker 1985, Bray 1983, Davidson 1980, Geddès 1983. Cités par S. Hodkinson: Bell and Limbrey (ed.) 1982, Braudel 1975.
Pour la Gaule antique et médiévale, une bibliographie est donné dans l'*Histoire de la France rurale* I (1975), période par période. De très nombreux articles paraissent périodiquement sur la question, souvent répétitifs. De même, peu d'éléments nouveaux ont été (par force) ajoutés à des ouvrages anciens. A consulter: J.F. Bladé, 'Essai sur l'histoire de la transhumance des Pyrénées françaises', *Bull. géogr. histor. et descr.* (1982); M. Brisebarre, *Bergers des Cévennes* (1978) (avec abondante bibliographie); N. Coulet, 'Sources et aspects de l'histoire de la transhumance des ovins en Provence au Bas Moyen-Âge' *Le monde alpin et rhodanien* (1978) 3-4, 213-247; P. Marres, *Les grandes Causses, étude de géographie humaine* (1935).
On trouvera également des éléments dans les histoires régionales: cf., par ex., *Histoire de l'Alsace rurale* (1983) 43 sq. Mais il n'existe aucun traitement synthétique de l'information, ni pour l'Antiquité ni pour les temps postérieurs.

2. On les trouvera recensées dans C. Jullian, *Histoire de la Gaule*, aux t. II et V, dans les chapitres concernant l'élevage, les industries (vêtements, peaux, etc.). On y ajoutera quelques références plus tardives concernant les manteaux aquitains (des Pyrénées): Greg. *Ep*. 2. 202; Paulin. *ad Aus. Ep*. 1. 246.

3. Cf. Jullian (n.2) V 239 et suiv. Les Lingons (ou les Leuques) étaient réputés pour leurs matelas de laine: Pline *N.H*. 8. 191-2 et *CIL* XIII 5444 (cf. P.-M. Duval, *Sources de l'Histoire de France* (1971) notice 134, p. 390). Les monuments figurés attestent l'emploi de la laine pour des tapis, des couvertures, des rideaux, des coussins, etc. Les peaux de boucs et de chèvres ont confectionné des sacs, des sacoches et probablement des outres, dont celles utilisées par les *utricularii*.

4. *Sagarii* (XII 1898, 1928, 1930, 2619, 3202, 4509; XIII 2008, 2010, 6744), *negotiator paenularius* (XIII 6366), *lanarii* (XII 4480-1). A quoi il faut ajouter les représentations ou inscriptions de *fullones*.

5. *CIL* XIII 7070 (= *Espérandieu* 5824): un *pecuarius* a été assassiné par son esclave à Mayence. Il est représenté avec un fouet, un chien et un troupeau de moutons.

6. La citation la plus reproduite est celle de Pline *N.H.* 11. 240: entre tous les fromages des provinces, les gourmets romains donnaient la palme au '*caseo Nemausensi, Lesurae Gabalicoque pagis*'. Beaucoup ont voulu voir là l'ancêtre du roquefort, redistribué par Nimes vers l'Italie. Seule occurence possible (quoique incertaine): Pline *N.H.* 11. 241: *Galliarum sapor medicamenti vim obtinet* (?).

7. Cf. *Espérandieu*, 40, 350, 626, 1044, 1240, 5824.

8. On se reportera aux études de Fr. Poplin, et notamment à: 'Remarques théoriques et pratiques sur les unités utilisées dans les études d'ostéologie quantitative', *IXe Congrès UISPP* (Nice, 1976), *Thèmes spécialisés: problèmes ethnologiques des vestiges osseux*, 124-141; 'Un problème d'ostéologie quantitative: calcul d'effectif initial d'après appariements' *Revue d'Archéométrie* 5 (1981) 159-165.

9. Résumons. Non seulement les conditions de fouille et la nature du terrain (propre ou non à conserver les os, y compris les plus petits, y compris les esquilles), mais aussi les procédures de comptage (nombre de restes par espèce, ou nombre minimum d'individus attesté par ces restes – étant entendu que deux tibias gauches attestent deux individus -, ou calcul par rapport au poids de viande – le poulet et le boeuf ne représentant pas la même capacité alimentaire). L'âge d'abattage des animaux compte aussi bien pour apprécier les goûts des consommateurs que pour estimer les autres services demandés au bétail (labours, charriages, reproduction, etc.). Une difficulté supplémentaire provient de l'identification: seuls quelques os permettent de distinguer la chèvre du mouton, ou le sanglier du porc – d'où, dans les tableaux, les classes 'mixtes' d'ovi-capridés (ou caprinés) et des suidés.

10. Parmi les plus récentes, Th. Poulain, 'La faune sauvage et domestique en France du néolithique à la fin de l'Âge du Fer', *La Préhistoire Française* II 104-115; *id.*, *L'étude des ossements animaux et son apport à l'archéologie* (1976). Ces deux contributions, fondées sur des résultats de fouilles parfois insuffisants en quantité d'ossements, sont surtout intéressantes pour la méthode. Une excellente synthèse de P. Méniel, *Contribution à l'histoire de l'élevage en Picardie*, *Rev. Arch. Pic., n° spécial* (1984). Ph. Columeau, *La chasse et l'élevage en Gaule méridionale d'après les vestiges osseux*, thèse inédite, Aix (1985).

11. Poulain (1976) 84 et suiv.

12. Ces chiffres sont fondés non sur les poids de boucherie actuels mais sur les rendements attestés au Moyen-Âge. Cf. Columeau (n.10) 19 et suiv.

13. Nous ne tiendrons pas compte de leur éventuelle prédominance à l'époque gallo-romaine en Vendée, dans les Charentes ou en Normandie: les analyses portent au total sur... 89 animaux!

14. Même impression en Picardie (Méniel) (n.10) si l'on calcule non en fonction du nombre de restes déterminés mais en fonction du poids de viande: sur 18 sites s'étageant du néolithique à la conquête, le boeuf est en tête dans 15 cas, et le porc dans 3; les ovi-capridés sont à la troisième place dans 17 cas sur 18.

15. Toujours pour le bétail, Pline *N.H.* 10. 53: les oies des Morins conduites jusqu'en Italie... à pattes!

16. Raison de plus pour souhaiter des analyses paléozoologiques sur des gisements médiévaux et surtout modernes!

17. Voir notamment H. Dubled, 'L'organisation de la boucherie et la consommation à Carpentras au XVe s.', *Bull. Provence Hist.* (1967); L. Stouff, 'La viande, ravitaillement et consommation à Carpentras au XVe s.', *Annales* 24 (1969) 1431-1448; *id.*, *Ravitaillement et alimentation en Provence aux XIVe et XVe s.* G. Désert, 'Viandes et poissons dans l'alimentation des Français au milieu du XIXe s.', *Annales* 30 (1975) 519-536.

18. Brisebarre (n.1) 57 et suiv. Les premiers règlements écrits sur la transhumance datent de la fin du XIIIe s.

19. Par exemple, pour les sites des Vosges, Fr. Petry, *Histoire de l'Alsace rurale* (1983) 60 et suiv., manifeste des réserves par rapport à J. Harmand, 'Le groupe des Viehwege vosgiens: une zone d'élevage en basse montagne', *Actes du Colloque Elevage et Vie pastorale* (Clermont-Ferrand, 1982).

20. Les références aux mondes orientaux offrent à l'ethnologue, à l'archéologue et à l'historien nombre de thèmes de réflexion. Ainsi, en Chine (au moins septentrionale), le mouton semble avoir joué un rôle important à l'Âge du Bronze (J. Gernet, 'Comportements en Chine archaïque', *Annales* 7 (1952) 31-38) puis fut totalement éliminé. Des textes orientaux font la différence entre le mouton qui a besoin d'être 'sur-protégé' (le berger et ses chiens font tout: la route, la défense, les pacages, etc: 'les moutons ont un caractére doux et ne savent pas se défendre': Nong Chou, trad. J. Gernet, 1950, alors que 'le boeuf a même souffle et même sang que l'homme') et, par exemple, le buffle qui, en troupeau, défendra son gardien contre le tigre et se préservera lui-même, au point que ce sont les enfants qui sont commis à la garde.

L'évolution tient sans doute moins à ces différences de 'caractère' qu'à une différenciation d'usage. Alors qu'au départ les troupeaux étaient utilisés pour la préparation des terrains (piétinement pour enterrer les graines semées à la volée), le dépiquage et la séparation des grains et de la paille aprés la moisson, ils furent ensuite réservés au trait (est-ce parce qu'on cherche, comme l'écrit O. Lattimore, 'La civilisation mère de barbarie', *Annales* 17 (1962) 105, à 'ne pas nourrir des animaux sur une terre qui pouvait être cultivée par l'homme'?). La Chine ignora longtemps la laine et le lait. Du coup, toutes nos références politico- idéologiques sur le pasteur (Caïn l'agriculteur a tué Abel le pasteur; Agamemnon était 'pasteur des peuples'; la parabole évangélique sur le bon pasteur, qui a inspiré tant de décors de tombes ou de sarcophages) ne se retrouvent pas làbas. Cf. A.G. Haudricourt, 'Domestication des animaux, culture des plantes et traitement d'autrui', *L'Homme* 2 (1962) 40-50.

21. J.-J. Hatt, *Rev. Arch. de l'Est* 31 (1980) 261-269.

22. W. Dehn, 'Transhumance in der Westlichen Späthallstattkultur', *Arch. Korrespondenzblatt* 2 (1972) 125 sq; *id.*, 'Le rôle de la transhumance pour les migrations des peuples celtiques', communication orale au *Congrès de l'UISPP*, Nice, 1976. Pour W. Dehn, une transhumance 'horizontale' menait les troupeaux hallstattiens (si on ose dire) depuis l'Allemagne Centrale vers la Bourgogne, puis, éventuellement vers le sud de la France. Ces pratiques auraient pu 'préparer' sinon des migrations du moins des infiltrations progressives vers le Midi. D'autre part, W. Dehn insiste sur la condition du berger qui, dans l'Antiquité, loin d'être un personnage en marge, jouirait de la condition honorable de 'conducteur en chef' des troupeaux.

23. C'est, par exemple, la thèse développée par A. Soutou pour les relations entre la Méditerranée et le Massif Central, dès le VIe siècle avant J.-C.: A. Soutou, 'La draille d'Aubrac et la progression hallstattienne dans le sud du Massif Central', *C. Lig. Pr. et Arch.* 9 (1959) 37-51.

24. Ainsi, la voie reliant Nimes à Clermont-Ferrand, dite 'regordane', est rapprochée du mot occitan 'regor', désignant l'agneau né pendant la période d'estivage, en dehors des périodes normales d'agnelage: il s'agirait donc de la draille par laquelle les 'regords' descendaient avant la fin de la période d'estivage: Brisebarre (n.1) 98.

25. Dans le Languedoc oriental, par exemple, les tumulus ont longtemps été considérés comme 'le témoignage d'invasions celtiques ou hallstattiennes, hordes de pasteurs venus des bords du Rhin par les crêtes qui bordent la vallée du Rhône ou par les drailles de l'intérieur du Massif Central, en marche vers les Pyrénées et la Péninsule Ibérique à travers les garrigues languedociennes': citation extraite de B. Dedet, 'Les tombes du Languedoc oriental au premier Âge du Fer', *RAN* 12 (1979) 9-42, lequel conclut à l'inanité d'une distinction entre un Languedoc 'des garrigues' (les pasteurs) et un Languedoc 'des plaines'. A l'inverse, lorsque nous cherchions à établir le recensement des *oppida* de l'Âge du Fer dans les Alpes Maritimes, nous nous sommes aperçu que des parcs à moutons modernes avaient été considérés (notamment par les prospecteurs aériens) comme des *oppida* protohistoriques.

26. Cf. J. Guilaine et J. Courtin, *l'Elevage en Méditerranée* (1977); D. Geddès, *De la chasse au troupeau en Méditerranée occidentale* (*Archives d'écologie préhistorique*, V, 1980); J. Guilaine *et alii, L'Abri Jean Cros* (1979) 397 sq. Une fouille particulièrement intéressante en Ardèche, celle de la Baume de Ronze à Orgnac, à propos de laquelle M.A. Beeching a bien voulu nous communiquer les faits suivants: 'Des contrôles à plusieurs endroits de cette énorme cavité ont montré que durant plus d'un millénaire d'occupation (entre 3800 et 2500 avant notre ère environ) il n'y avait aucun habitat vrai mais toujours présence d'animaux parqués. La fonction unique était donc bien celle de *bergerie en grotte*, sans doute avec cloisonnements internes (aire de 20 m² à 30 m², circulaires). Une recherche m'a montré que c'était sans doute le cas à cette époque pour toutes les cavités de la région Cèze-Ardèche; les fouilles de mon collègue R.Gilles et les études sédimentologiques de J.E. Brochier sont venues confirmer ces observations. J'ai retrouvé ensuite les emplacements des villages en plein air, à la périphérie de cette zone de garrigue assez inhospitalière. Je postule une situation de complément entre ces deux types d'installations: les sites de plein air étant les grands villages de base et les grottes des bergeries probablement saisonnières. Par ailleurs, la typologie céramique permet de préciser la notion de territoire. Les troupeaux étaient bien sûr gardés par quelques bergers, ce qui explique la présence des nombreux foyers culinaires ou calorifères, mais il n'y a pas de sol d'habitat au sens strict. Il semblerait donc s'agir d'un pastoralisme affirmé entraînant des activités spécialisées, mais des déplacements assez courts (10 à 20 km) sans doute saisonniers.'

Voir aussi: R. Gilles, 'Fouilles de sauvetage à la grotte de St-Marcel (Ardèche)', *Actes du Colloque sur le néolithique ancien méditerranéen* (*Arch. en Languedoc*, 1981) 267 sq.; D. Geddès, 'Les débuts de l'élevage dans la vallée de l'Aude' *Bull Soc Préhist Franc.* (1981) 370-378; J.E. Brochier, 'Combustion et parcage des herbivores domestiques', *Bull. Soc. Préhist. Franc.* (1983) 143-144; J. Maury, 'Le néolithique ancien des Grandes Causses', *Actes du Colloque sur le néolithique ancien méditeranéen* (*Arch. en Languedoc*, 1981) 261-264; J. Guilaine *et alii*, *Carsac* (1986) (notamment 255 sq). Etc. Merci à J. Guilaine d'avoir synthétisé pour nous les recherches récentes.

27. Cf., en dernier lieu, J. Guilaine *et alii*, *Carsac* (n.26) 255: 'évolution peu sensible' C'est également, on l'a vu, la conclusion de Ph. Columeau. En Picardie, en revanche, P. Meniel (n.10) 41, indique que 'le boeuf voit son importance numérique diminuer constamment, particulièrement entre le néolithique et l'Âge du Fer; le porc, au contraire, devient de plus en plus abondant; le mouton et la chèvre se

maintiennent, les proportions présentent des fluctuations de faible amplitude' – mais la Picardie n'est pas une région de pastoralisme.

28. Arbos (n.1). C'est dire que les causes de la transhumance ne se limitent pas à des phénomènes climatiques. Arbos cite un exemple 'exceptionnel' de transhumance échappant aux contraintes liées à l'agriculture: celle qui relie la Crau aux Alpes. Notons que c'est aussi le seul exemple cité par un texte antique: Pline (cf. *supra*).

ANIMAL BREEDING ON THE DANUBE

Regarding the animal husbandry of the European provinces of the Roman Empire there are a lot of scattered data known. Since the last decades of the 19th century several smaller or larger works (up to the level of doctoral dissertations) have dealt with the animal remains of Roman provincial sites. Lately, four monographs have evaluated the animal bone assemblages of one or several Roman provincial sites giving a detailed economic historical analysis about them and even discussing their wider interactions.

In sharp contrast with the above facts, we hardly know anything about the animal husbandry of Italy, the centre of the Empire, from biological sources, apart from the thorough report of Blanc and Blanc (1958-9, 7-57) on the animal remains of the Niger Lapis on the Forum of Rome and some smaller publications by Riedel (1974, 53-88, 1976, 105-122). This is in spite of the fact that O. Keller published his standard and still usable two-volume work *Die antike Tierwelt* (Keller, 1909, 1913) that contained an immense amount of written data and artistic representations concerning the domestic and wild animals of the antiquity. Thus the strange situation has developed that nowadays we know essentially more about the animal husbandry and domestic animals of the provinces of the Roman Empire than about those of the centre of the Empire itself and we are painstakingly trying to identify the domestic breeds found and osteologically described in the course of provincial excavations with those in the descriptions of antique authors.

The situation outlined above can only have one explanation: in Italy excavations have yielded archeological finds in such large quantities and high artistic quality that the excavators were not interested in such banal finds as animal bones for a long time. Nowadays, however, when the interest of archaeologists has turned towards economic historical questions the collection of animal remains has already started at Italian excavations, too, and their study can easily bring surprising results.

In the Roman Danube provinces one can study the interactions of two animal husbandry types. One was the primitive type of animal husbandry of prehistoric character of the local inhabitants, the other was the highly developed animal husbandry of the Romans that was in fact a conscious animal breeding with well-planned breeding selection, which had the aim of increasing productivity. It involved appropriate feeding in quantitative and qualitative respects, and even had a special basic literature in the background. There could not have been any doubt which of the two animal husbandries would exert more influence on the other. In fact, Roman animal husbandry transformed the Danubian one totally, and its influence can be seen even in remote villages.

First of all the Romans introduced several new domestic animal species into the Danubian area just as into other provinces. However, their influence was particularly strong in Pannonia because this region was quite close to Italy, the heart of the Empire. Ass, cat, camel, goose, duck and pigeon were introduced by the Romans into the Danubian provinces increasing the number of domestic animal species which occur by about 50 per cent. They took peafowl and guinea fowl to other parts of the Roman Empire, and it is only a question of time until one identifies their remains in the animal bone samples of a Roman site in one of the Danubian provinces. In this way willingly or unwillingly they made the domestic fauna more colourful.

Then they introduced a lot of highly productive and sometimes exotic breeds of almost every domestic species. First of all they selected (probably several breeds of) their large longhorn cattle (Bökönyi, 1974, 128, 1984a, 23f). They, or more exactly their auxiliaries, introduced humped cattle (zebu) whose representation can be seen on a Roman tombstone in the Museum of Plovdiv, Bulgaria (Bökönyi, 1974, 130, Fig. 27). A large sheep breed and a probable oriental one, belonging to the *Zackel* group, also arrived with them (Bökönyi, 1984a, 43). Then came the large Roman military horses (Hilzheimer, 1924, 151, Boessneck, 1957, 108, 1958, 68 f, Habermehl, 1957, 84 fit Herre, 1958, 35, Bökönyi 1974, 263 ff, 1984a, 81), at least five dog breed from lap dog to greyhound (Hornberger, 1970, 113, Bökönyi, 1974, 323 ff, 1984a, 66 ff) and a large breed of chicken (Gandert, 1953, 23, Schrolizer, 1961, 21, Piehler, 1976, 75, Bökönyi, 1984a, 93 f), which ends the list of the breeds introduced.

With the help of these newly introduced breeds, which were products of a long selective breeding practice, they transformed the local domestic stock in a fairly short time. As a result, the size of individual animals started growing. It is true, there remained small aboriginal animals, too, but the average size of, for example, cattle or sheep increased by about 16-17 cm. This size increase was not the same everywhere. For instance, in the town of Gorsium lying in the *hinterland* it was more evident than in the castrum of Intercisa on the Danubian limes (as can be seen in Fig. 1, in comparison to the cattle of the Celtic oppidum of Manching).

Another characteristic of the animal breeding of the Danubian provinces was that several breeds lived together in practically every species. The best examples of this are shown by horses and dogs in this respect. Among horses there occurred two or three breeds, and among dogs five to six breeds that differ from each other in skull form, size and body proportions.

As I have already stressed (Bökönyi, 1984a, 118): 'Evidently the Roman Empire wished to utilize both the northern portion of the Balkan Peninsula and

Pannonia as an animal breeding territory. But while in the southern Balkans highly productive breeds lived at the time of the Roman conquest and even before it, and could be used to improve the local breeds at the beginning of conscious animal breeding even in Italy itself, the primitive breeds of low productivity from the northern Balkans and Pannonia could not have brought the expected profit. It is possible that the idea of exchanging these primitive types for improved Italian breeds arose, but there lay many obstacles along this particular path to improvement of breed. First, the number of high quality breeding stock in Italy was never great enough to supply all the needs of the provinces. Second, the speedy export of large masses of stud animals was impossible for technical and logistical reasons. Finally, the sudden change from traditional low productivity populations would have upset the local inhabitants and made them resistant to any change.

Given the above problems what strategy could the Romans follow? They undoubtedly exported many stud animals to Pannonia, significantly more for example, than to Germania... . The owners of the villa farms and Roman settlers either bred only purebred Italian animals or else (and this was more probable) used the purebred forms in cross-breeding with local domestic types. On the other hand, the aboriginal inhabitants of the province were probably impressed by the relatively enormous productivity of the Italian breeds. They, therefore, tried to procure them or at least take every opportunity to use them for cross-breeding with their own stock. It is not likely that this process took very much time. Advanced Italian forms of domestic animals occurred by the second century in Sarmatian settlements lying far from the *limes* border region. Similar phenomena occurred in other border areas of the Roman Empire. Such distribution was most characteristic of draught oxen which obviously could not influence the local cattle stock in terms of reproduction. This change clearly indicates, however, that if Barbarian peoples beyond the *limes* knew the great advantages of the new breeds, then the local inhabitants of the provinces themselves were certainly even more conscious of the possibilities provided by these advanced forms.

It is possible of course that these developments were not the results of centralized organizing work. The general picture of the animal husbandry system is very similar, however, to those found in the Median and Parthian Empires which does suggest some kind of central but not necessarily state organization. In these two latter empires, well organized conscious breeding was concerned first and foremost with horse because of climatic, geographic, and strategic reasons other species were of much less significance.'

This kind of organizing strategy for animal husbandry became so successful in the Danube provinces that the inhabitants were hardly forced to hunt in order to cover their animal protein needs. Hunting became more a sport or pastime and remained important only in castrum and military stations, which was connected with the somewhat freer way of life of soldiers living there (see Table 1).

On the top of animal husbandry there stood cattle as the main milk, meat, draught power and dung producer. They were followed by caprovines in the south and pigs in the north. Horses were frequent only in settlements where cavalry units were stationed. An important characteristic of the Roman animal husbandry along the Danube was the relatively high ratio of domestic fowl, and this can be considered as a first step toward modern domestic fauna. (Nowadays, around 80 per cent of animals slaughtered in Europe for meat are domestic birds, primarily chicken.)

The main meat producer was thus cattle. Beef constituted in Gorsium, for example, almost 77 per cent of the consumed meat (Bökönyi, 1984a, 102). Caprovines only produced 6.5 and pig 9.5 per cent of the meat quantity eaten in the town. Horse, donkey, dog and cat were probably not eaten, and the meat of domestic poultry was unimportant because chicken were mainly kept for their eggs.

In fact, far more pork was produced than the quantity which was consumed in animal breeding centres of the provinces. The pig samples of two Pannonian towns – Gorsium (Bökönyi, 1984a, 103) and Scarbantia (Bökönyi, in print) – prove this, through the age structure of the pigs. There were as many as 27 per cent full-grown (adult and mature) pigs in Gorsium and 58 per cent in Scarbantia. In contrast, the animal bone sample from the Neolithic and Early Bronze Age site of Sitagroi-Photolivos, Greece, (Bökönyi, 1984b, 37) has shown that a pig stock can be maintained on a constant level with a breeding stock of 10 per cent of the population and with one farrowing a year, if the number of offspring averaged five. This rate of reproduction does not exceed that for a wild pig population. Considering the much more sophisticated methods of the Roman animal breeding, the large number of full-grown animals at Gorsium and Scarbantia meant a more intense increase in reproduction.

In the sample of Gorsium there were 8 boars to 26 sows (some of the boars may have been castrated). According to classical Roman authors one boar to ten sows was the ideal ratio in Italy. In a simplified calculation, 21 out of the 27 adult pigs (27 is the percentage of pigs in Gorsium) can be assumed to be sows. According to Keller, seven to eight of the newborn piglets were saved. Thus seven such offspring may be included in this calculation. In addition, if a farrowing average of 1.5 is assumed in any given year (this was the average reproductive rate of primitive and improved Roman pigs as well), then the 21 sows could produce 220 piglets per year. In this case, even if thirty of the piglets died, the net increase in the stock was 190 individuals each year. Since the first farrowing can occur when the sow is about two years of age, the majority of the sample 20 sows which were killed in their subadult age could produce 140 more offspring. Altogether this means an increase of 330 per cent in the stock every year. If about 73 per cent of the pork was consumed at the settlement, almost 260 per cent of the 330 per cent increase still remained to be sold. This ratio was even larger in Scarbantia. This

Table 1. Fauna lists of four sites from Roman Pannonia: Gorsium (town), Intercisa (Castrum), Pilismarót (military watch-tower), Scarbantia (town).

	Gorsium	Intercisa	Pilismarót	Scarbantia
cattle	38.33	42.80	31.65	63.10
sheep/goat	21.17	16.59	21.97	10.51
pig	19.26	21.41	28.55	16.20
camel		0.03	—	—
horse	6.56	8.95	16.53	7.15
ass	0.07	0.14	—	—
cat	0.24	0.45	—	0.13
dog	9.37	4.59	2.20	1.45
fowl	5.00	5.04	—	1.46
domestic	97.65	95.55	78.58	97.89
wild	2.35	4.45	21.42	2.11

hypothesis is supported by the fact that the bones of an enormously large number of juvenile individuals are missing in these sites. That means that these animals were either driven away or killed and taken away in two halves, smoked or salted.

One may suppose that these young pigs were sold to the army mainly in the border areas but in the interior, too. In fact, the inhabitants mainly consumed the cheap beef and sold the more expensive pork to the army or perhaps exported it to Italy.

INSTITUTE OF ARCHAEOLOGY, BUDAPEST S. BÖKÖNYI

REFERENCES

Blanc, D.A., Blanc, A. C. 1958-59, 'Il bove della stipe votiva del Niger Lapis del Foro Romano', *Bull. di Paletn. Ital.* N.S. 13 (67-68) 7-57.

Bökönyi, S., 1974, *History of domestic mammals in Central and Eastern Europe.*

Bökönyi, S., 1984a, 'Animal husbandry and hunting in Tác-Gorsium. The vertebrate fauna of a Roman town in Pannonia', *Stud. Arch.*, 8.

Bökönyi, S., 1984b, 'Die Herkunft bzw. Herausbildung der Haustierfauna Südosteuropas und ihre Verbindungen mit Südwestasien', in Nobis, G. (ed.), *Der Beginn der Haustierhaltung in der "Alten Welt". Die Anfänge des Neolithikums vom Orient bis Nordeuropa*, 9, 24-43.

Boessneck, J., 1957, 'Die Tierknochen', in Krämer, W., *Cambodunumforschungen, 1953-I* (Materialh. d. Bayer. Vorgesch., 9) 103-116.

Boessneck, J., 1958, *Zur Entwicklung vor- und frühgeschichtlicher Haus- und Wildtiere Bayerns im Rahmen der gleichzeitigen Tierwelt Mitteleuropas.* (Stud. an vor. – u. frühgesch. Tierrest. Bayerns, II, München).

Boessneck, J. et al., 1971, *Die Tierknochenfunde aus dem Oppidum von Manching.* (Die Ausgrabungen von Manching, 6).

Gandert, O.-F., 1953, 'Zur Abstammungs- und Kulturgeschichte des Hausgeflügels, insbesondere des Haushuhns', *Wiss. Abh. d. Akad d. Landwirtsch. zu Berlin*, 6, I, 69-81.

Habermehl, K.A., 1957, 'Die Tierknochenfunde im römischen Lagersdorf Butzbach', *Saalburgjahrb.*, 16, 67-108.

Herre, W., 1958, 'Die geschichtliche Entwicklung der Haustierzüchtung', in Zorn W., *Tierzüchtungslehre.* 17-43.

Hilzheimer, M., 1924, 'Die im Saalburgmuseum aufbewahrtgen Tierreste aus römischer Zeit', *Saalburgjahrb.* 5, 105-158.

Hornberger, M., 1970, 'Gesamtbeurteilung der Tierknochenfunde aus der Stadt auf dem Magdalensberg in Kärnten (1948-1966)', *Kärtn. Museumsscrr.* 49, 1-144.

Keller, O., 1909, 1913, *Die antike Tierwelt*, I-II.

Piehler, W., 1976, *Die Knochenfunde aus dem spätrömischen Kastell Vemania.* Diss. München.

Riedel, A., 1974, 'I mammiferi domestici della grotta N.1745/4558 V.G. e di faune oloceniche minori', *Atti e Mem. d. Commiss. Grotte 'Eug. Boegan'* 13, 53-88.

Riedel, A., 1976, 'La fauna del castelliere degli elleri (Trieste)', *Atti d. Mus Civ. di Stor. Nat. Trieste* 19 2, 105-122.

Schweizer, W., 1961, 'Zur Frühgeschichte des Hausnuhns in Mitteleuropa', *Stud. an vor- u. frühgesch. Tierrest. Bayerns*, IX, 1-42.

LE PASTORALISME DANS L'AFRIQUE ANTIQUE

Le pastoralisme occupe une place de choix dans l'historiographie du Maghreb puisque c'est au maghrébin Ibn Khaldoun que l'on doit la célèbre description du genre de vie bédouin (Umran Badaoun), société pastorale contrainte à la mobilité pour survivre et toujours jalouse de son indépendance. La soif de grands espaces des nomades en ferait les adversaires naturels de l'Etat. Ainsi le conflit éternel qu'ils entretiennent avec les sédentaires va bien au-delà des contradictions entre éleveurs et agriculteurs, inhérentes à toute civilisation paysanne. Même si cette présentation de la pensée d'Ibn Khaldoun est en réalité le résultat de la vision réductrice qu'en ont donné les historiens occidentaux à la suite d'E. F. Gauthier,[1] il n'en reste pas moins que l'opposition entre nomades et sédentaires a été considérée comme une des clés de l'histoire du Maghreb. Pour la période antique, cette vision est favorisée par les associations antinomyques que sont d'une part la Civilisation, le monde des villes et Rome, d'autre part la Barbarie, le monde rural, les nomades et les montagnards, l'Afrique que les Romains avaient oubliée pour reprendre l'image de Ch. Courtois.[2] Rien, certes, n'arrive sans raison: son pessimisme et les ambiguïtés de sa pensée ont permis la lecture simplificatrice qui a été donnée d'Ibn Khaldoun; l'image réductrice qui a été donnée de l'histoire du Maghreb antique est pour une part justifiée par les sources. Mais les progrès de la pensée géographique et l'apport de l'ethnoarchéologie permettent de renouveler les interprétations longtemps proposées; la réflexion sur les sociétés pastorales menée par les anthropologues permet de dépasser l'opposition simpliste entre sociétés nomade et sédentaire en rendant mieux compte de leur complexité et surtout de leur valeur propre : elle revalorise l'image du nomade sans dévaloriser celle du sédentaire.

Nomadisme et idéologie

L'ethnographie antique du nomade a déjà été étudiée.[3] Mais il est utile d'en rappeler les passages les plus célèbres. Ainsi dans son fameux portrait de Massinissa, Polybe écrit:

> Mais son oeuvre la plus belle, la plus divine fut celle-ci: avant lui, la Numidie toute entière était stérile et l'on pensait que son sol ne pouvait donner de récoltes; or, le premier, avec ses seules ressources, il prouva qu'elle pouvait produire toutes espèces de fruits (36.4.16).

Strabon lui fait écho:

> C'est ce prince qui le premier civilisa les Numides et les façonna à la vie agricole en même temps qu'il les déshabituait du brigandage pour leur apprendre le métier de soldat. Jusque-là les Numides avaient offert ce spectacle étrange d'un peuple en possession de terres éminemment fertiles mais infestées de bêtes féroces, qui, au lieu d'exterminer celles-ci pour cultiver ensuite ses champs en toute sûreté, avait mieux aimé se livrer à un brigandage sans frein et abandonner la terre aux reptiles et aux bêtes féroces, se réduisant ainsi à mener une vie errante et nomade ni plus ni moins que les peuples qui y sont condamnés par la misère, l'aridité de leur sol et la rigueur de leur climat. C'est même là ce qui fait donner aux Masaesyli la dénomination particulière de Numides. Dans ce temps-là naturellement leur vie était des plus simple; ils mangeaient plus souvent des racines que de la viande, se nourrissant en outre de lait et de fromage (17.15).

Cinq siècles plus tard, dans le *Bellum Vandalicum*, Procope reprenait ce *topos*:

> Les Maures habitent été comme hiver dans des huttes où l'on respire à peine ... Ils couchent par terre ... Ils ne changent pas de vêtement selon les saisons ... Ils n'ont ni pain ni vin ni aucune des choses qui rendent la vie agréable. Le blé, l'épeautre, l'orge leur servent de nourriture; mais ils ne savent ni moudre les grains ni les faire cuire: ils les mangent tout crus à la façon des animaux.[4]

Il y a peu encore, l'historiographie présentait la pénétration romaine au Maghreb comme une avancée de la civilisation et de la vie urbaine, caractérisée par la mise en culture de terres vierges, le cantonnement et la limitation des tribus. Cet espace romanisé et civilisé était défendu par une frontière fortifiée, le *limes*.[5] Rome aurait ainsi même refoulé dans le Sahara septentrional des berbères nomades contraints de chercher ailleurs des terrains de parcours.[6] Dans un brève publication où il exprime de fortes réticences par rapport à elle, P. Salama résume ainsi la théorie classique de L. Leschi: 'La politique romaine soucieuse d'assurer une quiétude absolue à ses territoires de colonisation interdisait purement et simplement l'accès de son domaine "utile" aux migrations saisonnières. De ce mobile politico-économique aurait résulté l'édification du *limes* séparant les Romains et les Barbares.[7] Dans *Les Vandales et l'Afrique* C. Courtois s'était déjà élevé contre la surévaluation du danger nomade et avait proposé de rechercher le danger barbare à l'intérieur même de l'Empire, dans les zones montagneuses: 'La civilisation romaine s'était répandue à la manière des eaux. Elle avait envahi les plaines sans recouvrir des montagnes. Si bien que l'Afrique se présentait comme une sorte de puzzle dans lequel se juxtaposaient les fragments du monde berbère et ceux du monde romain.'[8] Intégrant ces thèses pour dégager les fondements géographiques de l'histoire de l'Islam maghrebin, Xavier de Planhol y distingue trois ensembles définis par l'histoire et la géographie: les plaines, les zones montagneuses et les steppes ou déserts.[9]

Ces théories ont été soumises à un certain nombre de critiques. Il n'est pas lieu ici d'en reprendre le détail. Je me contenterai d'en rappeler les deux aspects méthodologique et idéologique. Pour écrire l'histoire, on se sert de quelques récits mentionnant des événements militaires auxquels on rapporte une documentation épigraphique et archéologique. Or celle-ci ne vise souvent qu'à nourrir ou illustrer une théorie déjà prête: c'est ainsi qu'un axiome épigraphique considérera que la tombe d'un vétéran suffit à prouver la présence d'une unité militaire. Le colonel Baradez a bâti une théorie du *limes* fondée sur de remarquables découvertes et sur le postulat d'après lequel un habitat de hauteur en position stratégique était militaire par destination, une route militaire par fonction et que tout aménagement était interprétable en termes stratégiques. La mise en question de l'axiome archéologie, symétrique du précédent, présente d'ailleurs des implications que mettent en évidence les discussions qui ont eu lieu à propos de la signification des remparts: de manière convergente, P.A. Février et R. Rebuffat ont montré qu'ils avaient une valeur symbolique et que l'on ne pouvait les comprendre si l'on n'en voyait que la fonction stratégique.[11] M. Euzennat qui soutient l'opinion contraire, a parfaitement raison de voir dans ce point l'une des divergences fondamentales entre 'les savants singulièrement qualifiés de coloniaux ou tout au moins considérés comme académiques'[12] et les autres dont en définitive le point commun semble être de tenir compte de l'apport des sciences sociales à l'histoire. C'est dans cette perspective qu'il convient de replacer l'interrogation de P. Trousset à propos du *limes*: 'Peut-on créditer d'une efficacité militaire assurée, un obstacle qu'il était aisé de contourner et qui au lieu de barrer des passages naturels dans un pays pourtant riche en coupe-gorges, se déploie sur le piedmont saharien en avant des grandes trouées du Hodna et de l'Aurès? C'est finalement en termes de surveillance des courants de circulation ... qu'il convient de reprendre l'interprétation de cette frontière.'[13]

Dans ce débat, il ne faut pas raisonner en terme de 'avoir raison' ou 'avoir tort' car tout le monde aurait un peu tort et un peu raison. Il faut bien dire à quel point de vue on se place. Ainsi à côté de la symbolique des historiens, il y a place pour la sémiologie des géographes: ils nous rappelent que le *limes* est aussi une traduction de l'opposition fondamentale entre le fermé et l'ouvert. Rempart et *limes* sont à des échelles différentes des systèmes de clôture étrangers à la mentalité du nomade qui au fermé préfère l'ouvert et dans la vie domestique, la tente à la maison.

Les acquis de la recherche récente (de la philologie à l'archéologie)

Contrairement à ce que l'on pourrait croire, même dans les disciplines philologiques, tout n'a pas été dit. Ainsi, pour ce qui nous intéresse ici, deux points sont à retenir dans les recherches de Denis Roques sur Synésios de Cyrène. Le premier est une contribution à la dénonciation du mythe d'une Cyrénaïque décadente économiquement à l'époque impériale: 'Les discours complaisamment ressassés depuis la Renaissance jusqu'à nos jours, dans la lignée de Strabon, Pline et Solin, sur la disparition du Silphion en Cyrénaïque dès les débuts du Haut-Empire' doivent être abolis.[14] Le second appelle à mettre en question une des 'grandes constantes' de l'histoire du Maghreb: 'La mise en évidence de la migration mazique et asturienne *ca* 400 invite à corriger une certaine vision de l'histoire du nord de l'Afrique. Les théories de E.F. Gautier et de C. Courtois sur les nomades chameliers venus de l'Est africain sous la pression d'un "mystérieux appel" ont habitué les esprits à n'envisager de mouvement de population ou de troupes que de l'est vers l'ouest ... Or la migration des Maziques et des Ausuriens *ca* 400 témoigne d'un ample mouvement inverse du Maghreb vers l'Est africain et même le Proche-Orient.'[15]

A la suite des recherches menées sur l'anthropologie de la Grèce antique surtout et à un moindre degré de Rome, on s'est aperçu que l'ethnographie antique fonctionnait sur des stéréotypes et qu'il fallait bien se garder de prendre les textes à la lettre lorsqu'ils décrivaient le Barbare nomade comme l'antithèse absolue du Romain civilisé.[16] J. Desanges vient de montrer que l'on pouvait étendre ce type de réflexion à un domaine particulièrement important pour notre propos, l'histoire du climat: une vieille tradition grecque que Polybe avait combattue affirmait la sécheresse et la stérilité de l'Afrique entière: après Strabon s'y substitue celle de la fertilité de l'Extrême-Occident. Les discours antiques sur le climat nous renseignent plus sur la pensée antique que sur le climat lui-même.[17] Nous mesurons ainsi les difficultés à évaluer la part de vérité contenue dans les récits 'légendaires'. Citons en un exemple significatif: P. Trousset voyait dans la description par Hérodote (IV,183) des boeufs qui paissent à reculons l'application aux Garamantes d'un 'fragment de discours parfaitement mythique: ils s'agit des boeufs d'Apollon dérobés par le jeune Hermès qui les fait paître à reculons pour tromper ses poursuivants.'[18] Il n'a certainement pas tort mais cette vérité mythique doit être acceptée comme telle à côté de la vérité réaliste rappelée par G. Camps qui observe qu'Hérodote tire la légende des boeufs opisthonomes des animaux aux cornes flottantes des peintures rupestres du Tassili n'Ajjer.[19]

Au Proche et au Moyen-Orient, sous l'influence de l'ethno-archéologie, la recherche pré- et proto-historique a renoncé à établir des chaînes causales simples entre agriculture, sédentarisation et vie en agglomération: l'agriculture n'implique pas un habitat sédentaire; de la présence d'un habitat sédentaire aggloméré, l'archéologue ne peut induire la certitude de pratiques agricoles.[20] En Afrique du Nord, les recherches sur l'occupation du sol tendent à montrer qu'il n'y a pas de coïncidence absolue entre la pénétration romaine, la sédentarisation agricole et l'urbanisation; ni dans le temps: sédentarisation et urbanisation sont antérieures à Rome; ni dans l'espace: la carte des installations sédentaires ne se calque pas sur celle du territoire contrôlé par Rome.

Il est démontré qu'au Maghreb le *limes* n'est pas la limite entre vies nomade et sédentaire. Pour ce qui est du Maroc, le point a été fait par E. Frezouls dans un article consacré à la réfutation d'une hypothèse reprise par une historienne anglaise, M. Sigman; interprétant l'histoire ancienne du Maghreb selon la problématique – bien désuette! – de la réussite ou de l'échec, celle-ci proposait comme clé 'le conflit sans cesse renaissant, jusqu'à ce que les occupants lâchent prise, entre la vocation nomade ou "semi-nomade", des provinciaux et la vie sédentaire implantée dans les centres romains ou romanisés, et que Rome aurait voulu généraliser dans le territoire annexé.'[21] A l'est, les recherches de J. Baradez sur le *fossatum* africain et celles de P. Trousset sur la partie tunisienne du *limes tripolitanus* ont montré que cet ouvrage n'était pas seulement une ligne de défense contre les nomades mais qu'en arrière de celle-ci se développait un semis de fermes fortifiées assurant la mise en culture du pays. Le *limes* apparaît maintenant plus comme un moyen de contrôler les déplacements des nomades que comme une barrière. La mise en culture était permise par une série d'ouvrages de petite hydraulique dont J. Birebent a montré l'importance et l'efficacité dans *Aquae romanae.*[22] Même situation dans le désert de Libye, où R. Rebuffat tire de ses prospections le tableau suivant de la 'civilisation des oueds': 'de petites villes côtières, dont quelques ports; une frange côtière occupée de façon dense; des vallées le long desquelles se sont développés des chapelets d'exploitations rurales; un réseau de pistes organisant et vivifiant le tout, et assurant les communications avec les régions limitrophes lointaines. Cette civilisation libyque conserve sa langue, et ses divinités, et sa vocation pastorale; elle a certainement subi l'influence punique en ce qui concerne le monde des images, et elle profite de la paix romaine, de l'organisation des communications et d'un recours systématique aux techniques hydrauliques et agraires.'[23] Ainsi, l'idée que l'on se fait maintenant du *limes* privilégie les facteurs humains de son existence et il n'est plus possible de l'expliquer par les seules notions physiques: la notion de frontière climatique, elle-même, apparaît comme un dernier avatar de l'idée de frontière naturelle et s'efface devant le 'constat d'une certaine ambivalence écologique autorisant, grâce au contrôle de l'eau, diverses combinaisons possibles entre vie sédentaire et semi-nomade.'[24]

Pour ce qui est de l'aspect chronologique du problème, deux choses sont à distinguer. A la question, l'agriculture a-t-elle été introduite depuis le Proche Orient par les Puniques et depuis l'Italie par les Romains, il faut bien entendu répondre par la négative. C'est en effet l'un des grands intérêts de l'ouvrage, que G. Camps publia en 1960, que d'avoir montré l'ancienneté des paysanneries sédentaires africaines, vérifiant ainsi l'une de ce qu'il a appelé les Vérités d'Hérodote; ce dernier opposait: 'la Libye orientale (où) habitent les nomades, (qui) est basse et sablonneuse jusqu'au fleuve Triton, et celle à l'occident de ce fleuve, habitée par les cultivateurs (qui) est très montagneuse, très boisée.'[25] Il est bien entendu que l'agriculture n'a pas été introduite par les Romains dans les oasis du désert. Mais c'est une autre question que de penser que Rome n'a pas étendu les limites de l'habitat sédentaire et de l'agriculture dans la zone steppique. Force est de reconnaître que dans la zone où se déroula le conflit avec les Numides entraînés par Tacfarinas on n'a pas trouvé de restes sûrs d'un habitat proto-historique. Sur le terrain, l'état actuel des connaissances est caractérisé par des ambiguités que seule une étude archéologique attentive permettrait de lever. Un exemple célébre illustre cette situation: en 1946, les ingénieurs du service cartographique découvraient les 'quadrillages de Tazbent' qui parurent – et c'est l'interprétation qu'en avait retenu G. Camps – appartenir à un système agraire et hydraulique cohérent, pré-romain, et même antérieur à Massinissa. En fait, il n'est pas sûr que l'ancienne interprétation des quadrillages de Tazbent ne doive pas être révisée: il s'agirait de systèmes d'enclos, de parcs utilisés par des éleveurs probablement nomades ou transhumants, ce qui n'a rien d'étonnant puisque l'on sait que le mouton a été domestiqué au Maghreb dès le Ve millénaire. En fait la région de Tebessa se trouve à la limite des deux zones méridiennes dont parlait Hérodote.

Dans un article où je m'élevais contre le thème de l'opposition de la plaine et de la montagne utilisé dans l'historiographie de l'Afrique antique, puis dans mes recherches sur les campagnes de *Caesarea* de Maurétanie,[26] j'ai traité du probléme du peuplement des zones montagneuses. Depuis, peu de travaux sont arrivés à terme, en dehors de la thèse de J. Peyras sur le *Tell nord-est tunisien.*[27] Bien qu'il m'ait reproché de 'minimiser trop le contraste entre la Maurétanie et les provinces orientales de l'Afrique', Cl. Lepelley exprime des opinions analogues dans son ouvrage consacré aux *Cités de l'Afrique romaine au Bas-Empire.*[28] Plus en effet que d'insister sur une différentiation régionale accusée par la documentation littéraire et épigraphique plus abondante à l'Est qu'à l'Ouest, il me semblait essentiel de dissocier les notions d'occupation du sol et d'occupation romaine. En Maurétanie Césarienne, l'occupation du sol diffère de nature entre les régions occupées et organisées à la romaine avec leurs centres urbains et leurs réseaux de *villae*, que la prospection autour de *Caesarea* m'a aidé à définir, et les territoires tribaux dont l'habitat est constitué de manière différente.[29] Xavier de Planhol et à sa suite R.I. Lawless ont eu parfaitement raison d'attirer l'attention sur l'occupation humaine de la 'frange interne du Maghreb, s'étendant du Haut et de l'Anti-Atlas du Maroc, à travers l'Atlas saharien et l'Aurès en Algérie, jusqu'aux djebels Matmata et Ouderna en Tunisie et jusqu'au djebel Nefousa en Tripolitaine ... Il semble que la population qui vivait dans les villages de pierres sèches ait été composée de paysans pasteurs pratiquant une agriculture d'irrigation et l'élevage du petit bétail.' Il est parfaitement admissible que 'cette civilisation paysanne ait réellement donné à la zone présaharienne du Maghreb un certain degré d'unité et une certaine individualité jusqu'à l'époque médiévale' et qu'il en résulte que l'opposition classique entre Tell et Sahara soit simpliste. Mais R.I. Lawless commet une grave erreur en affirmant que 'par contraste, les montagnes littorales restèrent à peu près vides pendant cette période et furent occupées par des communautés dont l'économie était beaucoup moins avancée et le niveau de vie très faible.[30] A sa base, on retrouve le présupposé qui avait longtemps brouillé la discussion sur le recul du contrôle romain en Afrique à la fin de l'Antiquité: l'identification entre présence romaine et occupation du sol,[31] forme politique et forme géographique d'occupation du sol.

Dans les campagnes de l'Empire romain, voisinnent des formes d'organisation de l'espace rural parfaitement hétérogènes qui n'obéissent à aucun déterminisme physique ou climatique. J'ai cherché à le montrer dans l'étude de la zone montagneuse sur les bords de laquelle fut fondée *Caesarea* de Maurétanie, ville romaine, capitale d'un royaume puis d'une province. A une campagne définie par le réseau des *villae* correspondant à l'aire d'influence de la ville s'opposait une campagne non romanisée définie négativement par l'absence de *villae*, dualité qui recoupe et réactualise l'opposition culturelle ressentie et exprimée par les Anciens entre monde urbain, civilisé et romain et monde barbare indigène. A l'autre bout de l'Afrique, en Cyrénaïque, sur les marges de la *chora* de Cyrène, A. Laronde constate que, de manière analogue, aucun déterminisme naturel, ne rend compte de l'imbrication constatée des espaces occupés par les sédentaires ou laissés aux nomades.[32]

L'apport des études géographiques

Jusqu'à une date récente, les géographes qui s'intéressaient aux questions d'occupation du sol dans l'Antiquité étaient des spécialistes de géographie humaine.[33] Or depuis quelques années ils reçoivent le renfort de pédologues et de morphologues du Quaternaire récent utilisant les vestiges antiques pour la datation des phénomènes de sédimentation ou d'érosion et s'interrogeant sur l'origine éventuellement anthropique des phénomènes; cette évolution est relativement nouvelle, en France au moins.[34] Quelle est, dans la morphologie de l'holocène récent, la part de l'homme et celle de la nature? Trois solutions sont possibles: le climat est resté stable à l'échelle historique et la désertification est d'origine anthropique; le désertification est liée à une pulsion climatique vers l'aridité; elle est le résultat d'une tendance à l'aridification aggravée par les activités humaines. C'est là qu'intervient le nomadisme: adaptation de l'homme au désert, qu'il s'agisse de populations évoluant sur place ou de groupes rejetés dans le désert, le nomadisme correspond à un processus qui, enclenché, se nourrit de lui-même.

Le problème est complexe. Dans son étude de la morphogenése holocène de la région de Cheria,[35] J.L. Ballais explique ainsi la formation d'un dépôt fin contenant des tessons de sigillée et comblant un canal d'irrigation: 'c'est la colonisation romaine qui, en développant l'agriculture et l'élevage ovin et caprin a permis la destruction d'une forêt rendue fragile par l'assèchement post-capsien. Alors, les sols marron minces ont été emportés par le ruissellement diffus, ruinant la région.' Pour lui donc, 'à l'époque historique, le moteur principal des grandes phases de modification de l'environnement passe des variations climatiques à l'évolution des sociétés humaines.'[36] Le géographe va donc rechercher chez l'historien des hypothèses explicatives confortant ainsi celle du refoulement des indigènes dans des zones montagneuses: les révoltes s'expliqueraient par le cantonnement dans l'Aurès des tribus berbères; celui-ci, par la surcharge qu'il entraînerait, expliquerait la destruction des sols. Rien de cela n'est invraisemblable, mais il faut savoir qu'historien et géographe prennent ici comme prouvé par l'autre ce qui pour lui n'était qu'hypothétique: dans le cas présent, et je me garde bien de le reprocher au géographe, aucun archéologue n'a sérieusement refléchi ni sur les données archéologiques ni sur les indices chronologiques utilisés.

S'il est sûr en effet que l'homme et le climat ont des responsabilités à se partager, il est bien difficile d'évaluer le rôle de chacun. Dans une communication au Congrès de Grenoble de la commission Afrique du Nord du CTHS, M. Rouvillois montrait comment des études palynologiques effectuées par A. Brun sur des sédiments marins carottés et dragués à proximité des îles Keniss et Kerkennah permettent de mieux poser le problème de la steppisation en Tunisie depuis l'époque romaine. Ces études ont en effet permis de préciser les idées que l'on avait de l'évolution de la couverture végétale de la Tunisie septentrionale et centrale. Elles paraissent montrer une tendance à la péjoration climatique, la forêt reculant au profit de la steppe; cette évolution aurait nourri une crise agricole se traduisant par le recul des paysans devant les nomades dont l'économie était mieux adaptée au nouvel environnement. A partir du Ve siècle jusqu'au XIe siècle, on assiste au développement de l'armoise, alors que le chêne recule du VIe s. av. J.-C. au VIe s. ap. J.-C. L'extension de l'armoise paraît d'origine plus climatique qu'anthropique: 'Si cette plante est particulièrement appétée par les troupeaux, elle devrait reculer plutôt que se multiplier. Il faut donc envisager que cette extension puisse être d'origine essentiellement climatique, d'autant plus que par le passé, sans intervention humaine notable, à chaque période plus aride correspond une extension spectaculaire de la steppe à chenodiacée et armoise.' D'où l'hypothèse que 'le nomade a suivi l'armoise beaucoup plus qu'il ne l'aura suscitée sur son passage.'[37]

Pour une étude anthropologique de l'espace

On doit, je crois, approuver M. Laraoui lorsqu'il met en évidence les sous-entendus politiques des théories de Gautier, Gsell et Courtois sur le Sahara, les nomades et les montagnards. Mais lorsqu'il écrit: 'le problème du Maghreb saharien ne relève ni du domaine de la zoologie ni de celui de la climatologie; il est d'essence historique,'[38] l'opinion du lecteur est partagée entre l'approbation de la dernière proposition et le refus de la première car ce n'est pas pour des raisons moins politiques qu'il présente le Sahara comme 'un réduit où se réfugie une partie de la population autochtone qui se re-nomadise tout en ne pensant qu'à repartir à la première occasion vers le nord.'[39] La tendance actuelle est en effet d'envisager d'une toute autre manière les rapports entre nomades, montagnards et sédentaires des plaines et des oasis.

P. Salama se demandait 'si l'organisation militaire et économique de la bande au sud de l'Ouarsenis n'était pas pour faciliter les contacts nomades-sédentaires par la création de centres, de marchés, de points d'eau où le côté bénéfique du nomadisme, c'est-à-dire avant tout l'apport des laines, aurait pu se manifester' et il proposait de voir dans le système romain 'la matérialisation d'une politique de paix aux frontières, connue par ailleurs en Maurétanie Tingitane et dans la préfiguration du système français sur l'organisation rationnelle de l'achaba.'[40]Une position analogue prévaut chez la plupart des auteurs traitant maintenant de ce thème. M. Euzennat résumait ainsi en 1974 l'acquis de la recherche récente: 'Le but poursuivi n'était certainement pas d'atteindre pour s'y fixer, comme le voulaient J. Baradez et J. Despois, l'extrême limite des possibilités de culture et de vie sédentaire ... Tout se passe ... comme si l'on avait voulu contrôler les hommes plus que les terres, dans un pays où celles-ci, en elles-mêmes, n'avaient pas de valeur. Loin de s'attacher à ce refoulement systématique des tribus nomades, auquel tous les historiens depuis Gsell ont voulu, on ne sait trop pourquoi, attribuer une importance décisive, les Romains ont en effet d'abord cherché à les inclure dans leur avance.'[41] J.M. Lassère a parfaitement raison d'écrire: 'Aucun document épigraphique ou littéraire ne vient donner une existence concrète à l'image romantique de tribus affamées tentant, au Haut-Empire, de forcer désespérament un *limes* destiné à contrôler leurs mouvements, et non à isoler deux mondes aux rapports nécessaires.'[42]

Ainsi, à travers des oppositions entre chercheurs considérant l'autre comme académique, traditionnaliste ou moderniste et ne se reconnaissant évidemment pas dans ces qualifications simplificatrices, se fait jour une approche nouvelle des problèmes qui tend à les dissocier d'une histoire événementielle fondée sur les connaissances ponctuelles que fournit l'épigraphie et d'une géographie prisonnière d'une conception linéaire et militaire de la frontière. L'article de R. Rebuffat, 'Au-delà des camps romains d'Afrique mineure', en constitue une excellente illustration: 'il faut totalement renoncer à l'idée qu'une barrière ait été érigée, au-delà de laquelle ne se serait trouvé qu'un monde inconnu et hostile. Au-delà des défenses fixes, l'armée contrôlait un vaste glacis. Au-delà de ce glacis, elle possédait encore cette sorte de défense avancée et non négligeable que constitue le renseignement ... Ni la Libye intérieure ni Cidamus ni la côte marocaine jusqu'à Mogador ne sortent à ce moment du monde romain, quoiqu'avec d'importantes différences de degré dans cette permanence.'[43]

Cette évolution est donc, on le voit, le résultat d'une réflexion sur le territoire poursuivie en liaison étroite avec le débat portant sur le nomadisme pastoral. L'historien des sociétés antiques a tout à gagner d'une réflexion menée avec les géographes et les ethnologues. Or celle-ci a évolué depuis la parution des *Fondements géographiques de l'histoire de l'Islam* en 1968. L'un des concepts de base utilisés par Xavier de Planhol avait été celui de bédouinisation, conçu comme passage d'une économie agro-pastorale (culture irriguée et élevage transhumant) ou pastorale à déplacement court à une société fondée sur un élevage extensif à grand rayon, 'nomadisme chamelier belliqueux' dont, à la suite d'E.F. Gauthier, il datait l'apparition de la fin du IIIe siècle après J.-C.[44] Or le concept de bédouinisation est actuellement remis en question. Comme l'écrit J. Vignet-Zunz dans le recueil *Pastoral Production and Society*: 'La steppe, la ville, la plaine et la montagne irriguée sont interdépendantes; ... complexe dont la ville, entrepôt et principal centre de transformation et de consommation en matière agricole, est un lieu privilégié: pour être plus souvent incluse au sein des terroirs agricoles, elle n'est pas nécessairement la chose de la société agricole; terrain de rencontre des horizons les plus disparates, elle a pu préparer, favoriser les conditions de fusion conflictuelles des formes élaborées, de part et d'autre, à partir du pastoralisme et de l'irrigation.'[45]

Cette réflexion est indissociable de celle que conduit sur l'espace l'équipe 'écologie et anthropologie des sociétés pastorale' au travers d'un débat sur l'organisation territoriale et la notion de territoire dans les sociétés de pasteurs nomades et dont on trouve les échos dans la revue *Production pastorale et société*.[46] L'anthropologie conduit à renoncer à voir dans les modes de vie nomade et sédentaire deux formes successives d'exploitation du territoire: elles sont l'une et l'autre la conséquence de la sédentarisation agricole et chacune s'accompagne d'une conception différente du territoire: à l'espace rayonnant, construit 'en cercle concentriques autour de son grenier' de l'agriculteur sédentaire s'oppose l'espace itinérant des chasseurs cueilleurs.[47] L'opposition entre sociétés nomade et sédentaire procède de deux conceptions de l'espace liées à leurs productions de base. Celles qui définissent les formations sociales nomades se caractérisent par leur mobilité: 'les troupeaux et non les terres constituaient toujours la richesse essentielle des pasteurs transhumants et définissaient leur système de propriété.'[48] L'opposition entre sociétés nomade et sédentaire ne peut être réduite à celle du désordre à l'ordre: la guerre elle-même peut très bien être conçue comme l'expression violente de la complémentarité entre nomades et sédentaires.[49]

La recherche moderne s'oriente donc vers la mise en évidence des complémentarités. Ainsi utilisant les idées d'Owen Lattimore sur la dialectique des rapports entre barbares des steppes et Etats agro-marchands, J. Vignet-Zunz présente le bédouin comme entrainant l'inclusion du désert et de la steppe dans l'orbite des cités agro-marchandes. En fait ce n'est en rien une nouveauté médiévale: dés le second millénaire avant J.-C., 'on voit pleinement se réaliser surtout au Proche-Orient, une civilisation agro-pastorale *et* commerciale issue de la "révolution néolithique".'[50] L'histoire de l'Afrique antique peut parfaitement être renouvellée à la lumière de ces nouvelles recherches sur l'espace de contact entre sociétés nomade et sédentaire, comme l'ont proposé parallélement P. Trousset et Brent D. Shaw.[51]

Les limites d'une approche anthropologique des rapports entre Rome et le monde indigène dans le Maghreb antique

La présentation par A. Leroi-Gourhan de la guerre comme forme violente de complémentarité est parfaitement admissible pour un historien des sociétés et des économies antiques.[52] Mais la vision anthropologique n'est qu'un élément de la compréhension du processus historique et ne peut pas être substituée à l'analyse événementielle des conflits qui constituent la trame même de l'histoire. C'est du moins l'objection qui, à mon avis, doit être opposée aux chercheurs de langue anglaise[53] qui développent une vision continuiste de l'histoire de l'Afrique du Nord, étendant à l'Antiquité et à l'invasion romaine la critique qui a été adressée à l'invasion hillalienne. Elle conduit à confondre la présentation par M. Benabou de la politique romaine de cantonnement des tribus et de sédentarisation des nomades avec l'attitude simpliste d'historiens qui se sont contentés d'inverser l'historiographie coloniale en transformant en héros ceux qu'elle présentait comme des barbares, rebelles à la Civilisation, et en traitres ceux qui s'étaient laissé tenter par le genre de vie romain.[54]

L'image de l'opposition, ou de la différentiation entre Rome et le monde indigène que je retire des prospections que j'ai pu faire dans les régions littorales, va dans le sens des idées de M. Benabou. Je pense avoir apporté l'illustration concrète de la manière dont une ville romaine a mis en valeur et exploité une campagne dépendante à laquelle ne s'opposait pas foncièrement: le clivage entre ville romaine et campagne indigène ne passe pas entre le rempart urbain et les premiers établissement agricoles. La rupture entre monde romain et monde indigène ne peut être tracée à partir des courbes de niveaux séparant 'la plaine' et 'la montagne'. Il n'y pas une, mais deux campagnes: une campagne organisée par la ville et définie par le réseau des *villae* et une campagne indigène africaine où l'analyse archéologique met en évidence un habitat qui est le reflet d'autres structures économiques.[55] Cette rupture, identifiable dans le paysage archéologique, peut être mise en relation avec des données épigraphiques et littéraires évoquant épisodes militaires, prélévement fiscaux et prélèvement de terres.[56] Sur les marges de la *chora* de Cyrène, A. Laronde a fait des observations comparables à propos des rapports entre la zone d'habitat sédentaire et les espaces qui paraissent avoir été laissés aux nomades (cf supra n.32) Comme M. Benabou, je crois à l'existence d'une politique romaine de sédentarisation des tribus du Sud correspondant à la limitation des tribus du Nord. J'adhère parfaitement à la critique qu'il fait de l'interprétation par Brent Shaw de la guerre de Tacfarinas selon lequel il ne s'agirait pas de la révolte de nomades spoliés mais de la rébellions de bandes d'irréguliers: comme il l'écrit: 'le recours massif à l'ethnographie finit par évacuer l'histoire.'[57] Lisa Fentress, dans la réponse qu'elle adresse aux critiques de Brent Shaw sur son ouvrage *Numidia and the Roman Army*, relève que l'importance accordée par celui-ci à l'environnement et l'utilisation de l'anthropologie le rapproche des tenants de la vieille thèse de la 'permanence berbère'; Brent Shaw sous-estime les changements apportés par Rome, en particulier par l'introduction de la vie urbaine.[58]

Les formes d'occupation du sol et la géographie historique actuelle

Le recherche archéologique a longtemps admis des schémas oppositionnels simples: ville/campagne, plaine/montagne, nomades/sédentaires, Romains/-indigènes. Ces oppositions sont toujours plus ou moins acceptées par l'opinion publique dans les pays du Maghreb: elle continue à assimiler Rome et la colonisation européenne, les nomades, les montagnards et la résistance à Rome. Mais aussi bien en Libye avec en particulier l'UNESCO Libyan Valleys Survey, en Tunisie avec des recherches conduites par des chercheurs de l'INAA seuls ou en collaboration avec des missions archéologiques étrangères, au Maroc où la Direction des Antiquités pré-islamiques et islamiques organise des inventaires archéologiques, l'apport de connaissances nouvelles[59] appelle une réécriture de l'histoire de l'occupation du sol.

Celle-ci doit tenir compte de l'évolution de la géographie historique actuelle. Ces dernières années, la recherche a beaucoup insisté sur la complexité des sociétés et la relativité des valeurs et l'on ne peut plus se contenter de voir dans le montagnard et le nomade des barbares faméliques prêts à se jeter sur la Civilisation! Notre image des sociétés primitives a complétement changé sous l'influence démystificatrice d'anthropologues comme Marshall Sahlins dans *Age de pierre, âge d'abondance*. Actuellement se manifestent des réactions contre cette vision rénovée de l'âge d'or. Déjà depuis quelques années s'étaient multipliées les critiques contre le possibilisme de Lucien Febvre et l'idéologie antidéterministe dominante jusque dans les années 1970. Depuis, les géographes ont commencé à réévaluer leur conception des rapports entre l'homme et l'espace: rompant avec le sociologisme et l'économisme, la géographie a redécouvert les contraintes du milieu et de l'histoire.[60]

Je crois que, pour le moment, tout a été dit sur les aspects idéologiques de l'historiographie des nomades et des montagnards en Afrique du Nord et que, dans ce domaine, le débat prend un tour irritant de discours autour des mots. Mais la prospection de nouveaux sites n'est pas le seul moyen de relancer le débat. Il est nécessaire d'intégrer l'apport d'autres disciplines. J'ai déjà fait état de deux d'entre elles: la paléobotanique et la géomorphologie dont les données ont déjà été prises en compte. Il faut y ajouter l'archéozoologie qui, en Afrique du Nord, n'a encore attiré l'attention que des préhistoriens. Or comment étudier sérieusement le problème du pastoralisme sans une meilleure connaissance des espèces élevées? Cette lacune documentaire a eu pour conséquence le développement du mythe de l'introduction du chameau par les Romains à une époque relativement récente et la théorie des conséquences prétendument catastrophiques de cette introduction pour les frontières militaires romaines.[61] Par ailleurs il ne faut pas surestimer l'apport de ces disciplines: c'est en fait par l'interdisciplinarité que l'on pourra avancer. Ainsi on s'est rendu compte que la palynologie ne devait pas être pratiquée sans discernement: en 'milieu ouvert', les couches stratigraphiques sont rapidement 'polluées' par des pollens venant des couches supérieures; les pollens sont transportés par les agents athmosphériques, parfois sur des distances considérables; ainsi arrivent au Sahara des pollens d'espèces croissant dans les régions humides. La géomorphologie est elle-même dépendante des datations fournies par les archéologues, et des interprétations proposées par les historiens. Une note cosignée d'un géomorphologue, R. Paskoff, et de deux archéologues, H. Hurst et F. Rakob, sur la 'Position du niveau de la mer et le déplacement de la ligne de rivage à Carthage dans l'Antiquité',[62] illustre bien l'intérêt et les difficultés de ce genre d'étude.

> Le trait le plus original des changements du littoral à Carthage est lié à l'élargissement de la plage, qui s'est manifesté entre 400 av. J.-C. et 150 ap. J.-C. et probablement même plus tôt. Cette progradation de la côte est logiquement à mettre en relation avec un accroissement du transfert sableux littoral. Un tel phénomène peut s'expliquer par le développement considérable de l'agriculture du temps des Carthaginois et des Romains dans les plaines de la Tunisie du Nord. On est en droit de penser que des défrichements étendus favorisèrent l'érosion des sols qui augmenta la charge alluviale livrée à la mer par la Medjerda. A une date inconnue se produisit un renversement de tendance dans l'évolution géomorphologique du littoral. Sans doute la progradation s'arrêta-t-elle lorsque, après la fin du Monde antique, la substitution d'une agriculture extensive à une mise en valeur intensive, pour l'époque, de la terre fit décliner l'érosion des sols.

Ainsi s'ouvre pour l'historien la perspective d'un renouvellement de la question de l'occupation du sol par la mise en évidence de l'interférence entre l'évolution de la ligne de rivage et les systèmes agricoles associés à des formes politiques successives. Mais la vérification de ces hypothèses suppose l'évaluation des quantités de sédiments déposés pendant la période de progradation du littoral et considérés comme arrachés aux sols de l'intérieur, l'étude de la constitution concomitante des terrasses alluviales de la Medjerda et celle des zones amont d'où est sensé provenir le matériel. Elle suppose aussi une véritable connaissance du mode d'occupation du sol aux époques carthaginoise et romaine et des systèmes agricoles, autant de données dont ne sont en fait connues que des tendances très générales. Les historiens discutent toujours de la signification de l'anecdote de Caton montrant au Sénat une figue fraiche arrivée de Carthage pour justifier son célébre *delenda est Carthago*: voulait-il simplement montrer que Carthage était proche de Rome? ou bien la figue n'était-elle pas là pour représenter la production d'une agriculture arbustive concurente de celle de Rome et qu'il convenait de détruire afin de réserver aux Italiens une spécialisation fructueuse? Après la défaite de Carthage, les Romains coupent les arbres; simple geste de punition et de vengeance ou acte réfléchi visant à faire de l'Afrique le producteur obligé d'une production céréalière indispensable mais pour cette raison même très contrôlée et peu rentable? Leur destruction n'exclut d'ailleurs pas une rapide reconstitution des vergers pour le plus grand profit des Romains.

Une fois ces limites admises, de grands espoirs peuvent être fondés sur les pratiques interdisciplinaires. Ainsi, pour ce qui nous préoccupe ici – le rapport entre l'homme et le milieu naturel –, l'une des voies possibles est la quantification des colluvionnements descendus des hauteurs dans les vallées: la datation des couches grâce à l'archéologie et l'évaluation des masses sédimentaires pourra permettre de quantifier la proportion de sédiments dont la mise en mouvement peut être imputée à l'intervention humaine. Alors, peut-être verra-t-on plus clair dans le problème de la stabilisation romaine des terroirs: l'érosion dans les montagnes, la steppisation sur les plateaux ont-elles connues une pause à l'époque romaine? Observe-t-on une différence réelle entre les régions contrôlées par Rome ou restées en dehors de son Empire? Y-a-t-il une spécificité maghrébine ou les choses s'y passent-elles d'une manière analogue à celle observée dans le reste du bassin méditerranéen? Ou, en d'autres termes, quel est le rapport entre l'écosystème méditerranéen et le système politique et social romain?

UNIVERSITÉ DE PROVENCE PHILIPPE LEVEAU

NOTES

1. M. Laraoui, *L'Histoire du Maghreb. Un essai de synthèse* (1970) 202-203.

2. C. Courtois, *Les Vandales et l'Afrique* 1 (1955) titre du ch.2 de la 3e partie.

3. P. Trousset, 'L'image du nomade saharien dans l'historiographie antique' dans *Production pastorale et société* 11 (1980) 97-105.

4. Citation et traduction par Ch. Diehl, *L'Afrique byzantine* (1898) 309.

5. L'article de base est celui de L. Leschi, Rome et les nomades du Sahara central, dans *Travaux de l'Institut de Recherches Sahariennes* (1942) 47-62, = *Etudes d'épigraphie, d'archéologie et d'histoire africaines* (1957) 65-74. Cf également, C. Courtois, (n.2) 101.

6. Théorie due à S. Gsell, 'La Tripolitaine et la Sahara au IIIe siècle de notre ère', *Mémoires de l'Académie des Inscriptions* 43 (1926) 149-166. Elle continue à jouir d'une certaine faveur dans l'historiographie maghrébine actuelle pour des raisons qui ne sont pas étrangères aux *a priori* idéologiques: cf M. Laraoui, (n.1) 69 et 70: 'En définitive, comme c'est généralement le cas, l'intuition de Gsell semble être infiniment plus acceptable que les thèses ultérieures ...'

7. P. Salama, 'Un point d'eau du limes maurétanien', dans *Maghreb et Sahara, Etudes géographiques offertes à Jean Despois*, n° spécial d'*Acta geographica* (1973) 342.

8. C. Courtois, (n.2) 113.

9. X. de Planhol, *Les fondements géographiques de l'Histoire de l'Islam* (1968) 126.

10. L'historiographie de la révolte maure de l'époque d'Antonin le Pieux constitue un excellent exemple de ces problèmes liés au traitement de la documentation. Cette révolte a été en partie créée par S. Gsell et J. Carcopino dans un article classique où ils ont rassemblé des données militaires dont les rapports n'étaient pas forcément évidents; J. Baradez y a raccroché des données archéologiques discutables. Sur cette affaire, cf ma préface à l'ouvrage de N. Benseddik, *Les troupes auxiliaires de l'armée romaine en Maurétanie césarienne sous l'Empire* (1982) 7-8. De même R. Rebuffat, 'Enceintes urbaines et insécurité en Maurétanie Tingitane', *MEFRA* 86 (1974) 501-521 et plus récemment M. Christol des, 'L'armée, des provinces pannonienes et la pacification des révoltes maures sous Antonin le Pieux' *Antiquités Africaines* 17 (1981) 133-141, ont montré que l'on avait eu tendance à gonfler cette révolte.

11. R. Rebuffat, (n.10); P.A. Février, 'Enceinte et colonie', *Revue d'études ligures* (1969) 278.

12. M. Euzennat, 'Les troubles de Maurétanie', dans *CRAI* (1984) 373.

13. P. Trousset, 'Notes sur un type d'ouvrage linéaire du limes d'Afrique', *BCTH*, 17 B (1984) 391-392.

14. D. Roques, 'Synésios de Cyrène et le Silphion de Cyrénaïque', *REG* 97 (1984) 231.

15. *Id*., Synésios de Cyrène et les migrations berbères vers l'Orient (398-413), *CRAI* (1983) 675. G. Camps, *Les Berbères aux marges de l'histoire* (1980) évoquait au contraire 'une de ces pulsions qui, depuis la Préhistoire, conduit toujours au Maghreb des populations venues de l'Est' (p.125).

16. D'une manière générale, on se reportera à P. Briant, *Etats et pasteurs au Moyen-Orient ancien* (1982) et plus particulièrement pour l'Afrique à l'article de P. Trousset (n.3). Le très volumineux travail de Y. Dauge, *Le Barbare, recherches sur la conception romaine de la barbarie et de la civilisation* (1981), peut être complètement négligé sauf pour une présentation caricaturale de la pensée 'académique'.

17. J. Desanges, 'De Timée à Strabon, la polémique sur le climat de l'Afrique du Nord et ses effets', résumé d'une communication présentée au Congrès des Société Savantes (1985).

18. P. Trousset, (n.3) 99-100.

19. G. Camps, *Aux origines de la Berbérie, Massinissa ou les débuts de l'histoire* (*Libyca, arc., épigr.*, 8 (1960)) 17: Les Vérités d'Hérodote. G. Camps traite de cette autre vérité dans 'Pour une lecture naïve d'Hérodote – Les récits libyens (IV, 168-169)', dans *Histoire de l'historiographie* 7 (1985) 38-59. Je ne tiens pas ce raisonnement pour faire plaisir aux deux partisans de ces 'vérités' mais parce que je les crois complémentaires. C'est une constatation de beaucoup de scientifiques qu'il existe une 'discontinuité qui interdit au chercheur d'unifier totalement les résultats qu'il obtient sur le même objet à des échelles et avec des méthodes différentes', ce que Jean Hamburger caractérise par le concept de Césure (*La raison et la passion, réflexion sur les limites de la connaissance* (1984) 40).

20. Ainsi les habitants des premières agglomérations villageoises de la civilisation 'natoufienne' sont encore des chasseurs cueilleurs (cf J. Cauvin, *Les Premiers villages de Syrie Palestine du IXe au VIIe millénaire avant J.-C.* (1978).

21. E. Frézouls, 'Rome et la Maurétanie tingitane: un constat d'échec?', *Antiquités Africaines* 16 (1980) 65-93.

22. J. Baradez, *Fossatum Africae* (1949); P. Trousset, *Recherches sur le limes tripolitanus du chott el-Djerid à la frontière tunisienne* (1974); J. Birebent, *Aquae romanae* (1962)

23. R. Rebuffat, Recherches dans le désert de Libye, dans *CRAI* 1982 196-197. Les résultats des Libyan valleys survey de G.B. Jones et G.W. Barker sont parus dans les *Libyan Studies* 11-15 (1979-84)

24. P. Trousset, 'Limes et frontière climatique', *III Colloque sur l'histoire et l'archéologie de l'Afrique du Nord, Montpellier 1985* (1987), 55-84, *id.*, 'Les oasis présahariennes dans l'Antiquité: partage de l'eau et division du temps', *Antiquités Africaines*, 22 (1986) 163-193.

25. G. Camps, (n.19) 17-18

26. Ph. Leveau, 'L'opposition de la plaine et de la montagne dans l'historiographie de l'Afrique du nord antique', *Annales de Géographie* (1977) 201-205.

27. J. Peyras, *Le Tell Nord-Est Tunisien*, thèse pour le Doctorat d'Etat soutenue auprès de l'Université de Bordeaux III en 1984.

28. Cl. Lepelley, *Les cités de l'Afrique romaine* (1979) 52.

29. Ph. Leveau, *Caesarea de Maurétanie, une ville romaine d'Afrique et ses campagnes* (1984) 476 et n. 55, 483-485 (socio-topographie des campagnes), 488-489 (les deux campagnes et les deux Afriques).

30. R.I. Lawless, 'L'évolution du peuplement, de l'habitat et des paysages agraires du Maghreb', *Annales de Géographie* (1973) 451-452 et 464.

31. Ainsi Cl. Lepelley ((n.28) 51) approuve M. Benabou en le citant: 'ce que C. Courtois appelait Afrique abandonnée s'est maintenant singulièrement rétréci'. (M. Benabou, *La résistance africaine à la Romanisation* (1976) 239).

32. A. Laronde, 'Connaissance géographique et exploitation rurale de la chôra cyrénéenne pendant l'antiquité classique', *Bull. Assoc. Géogr. Franc.*, 499 (1984) 33-39.

33. Je pense en particulier à J. Despois que P. Salama appelle 'géohistorien du Maghreb' en exergue à l'article cité (n. 7) qu'il a écrit dans les *Mélanges* qui lui étaient offerts, ou encore à Xavier de Planhol.

34. On en trouvera une bonne présentation dans *La mobilité des paysages méditerranéens* (Hommage à Pierre Birot), Travaux II de la *Revue Géographique des Pyrénées et du Sud-Ouest* (1984) en particulier Troisième partie: Sensibilité des milieux et sociétés méditerranéennes. Pour la vallée du Chélif, il y aurait eu, pour un archéologue, un intérêt considérable à travailler avec le pédologue J. Boulaine, *Etude des sols des plaines du Chélif* (1957); j'ai essayé de tirer profit de ses indications dans mes 'Recherches historiques sur une région montagneuse de Maurétanie césarienne: des Tigava Castra à la mer' *MEFRA*, 89 (1977) 390-394 et 311.

35. J.L. Ballais, La morphogenése holocène de la région de Cheria (Némentchas, Algérie), *Actes du symposium sur les versants en pays méditerranéens* (1975) CEGERM, vol. 5, 127-130.

36. *Ibid.*, M. Sary rappelle que 'les romains ne sont pas les seuls responsables de la dégradation forestière. Ils ont repris la technique d'épandage de la crue des autochtones.' Sur ces problèmes, *cf. La mobilité des paysages* ... , (n.31) 323-330 (La pesée des responsabilités humaines dans le déclenchement de crises morphogéniques: le 'laboratoire maghrébin').

37. M. Rouvillois-Brigol, 'La steppisation en Tunisie depuis l'époque punique: déterminisme humain ou climatique?' *BCTH* 19B (1985) 215-224. M. Rouvillois-Brigol a fait une nouvelle communication à Montpellier en 1985 sur le sujet: 'Quelques remarques sur les variations de l'occupation du sol dans le Sud-Est algérien'. L'étude de A. Brun a été publiée dans *Pollen et Spores* (1984) (Etude palynologique des sédiments marins holocènes de 5000 BP à l'actuel dans le golf de Gabès (mer Pélagienne)).

38. M. Laraoui, (n.1) 70.

39. *Id.*, (n.1) 70.

40. P. Salama, (n.7) 343.

41. M. Euzennat, 'Recherches récentes sur la frontière d'Afrique (1964-1974)', *Studien zu den Militärgrenze Roms, II, Vorträge des 10. internationalen Limescongresses in der Germania Inferior* (1977) 441. (pour comprendre le 'pourquoi', on se reportera à M. Laraoui). Cf. également Cl. Lepelley, (n.28) 45.

42. J.M. Lassère, *Ubique Populus. Peuplement et mouvements de population dans l'Afrique romaine de la chute de Carthage à la fin de la dynastie des Sévères (146 a.C.-235 p.C.)* (1979) 361.

43. R. Rebuffat, 'Au-delà des camps romains d'Afrique Mineure', *ANRW*, 10, 2 (1982) 508-509.

44. X. de Planhol, (n.9) 131, renvoyant à E.F. Gautier, *Les siècles obscurs du Maghreb* (1927) 165-185.

45. J. Vignet-Zunz, 'A propos des Bédouins: une réévaluation des rapports "nomades-sédentaires"', *Pastoral production and Society* (1979) 477. On consultera également du même auteur: 'A propos de la Cyrénaïque: quelques reflexions sur "bédouin" et "bédouinisation"', *Cahier du CERM*, 133 (1977) n° 4 de la série, *Etudes sur les sociétés des pasteurs nomades*, 22-41; 'A propos de la bédouinisation en Afrique du Nord: déterminations culturelles et niveaux des techniques agricoles' *Techniques et cultures* (1983) (Actes de la Table Ronde 'technologie culturelle' (Ivry novembre 1982) 49-70).

46. Revue publiée par la Maison des Sciences de l'Homme à partir de l'hiver 1977-1978. après le colloque d'ou était sorti l'ouvrage cité (n.41). Cf. également X. de Planhol, 'Nomades et pasteurs XVI-XIX, A travers les colloques sur le nomadisme pastoral', *Revue Géographique de l'Est* (1979) 303 sq.

47. A. Leroi-Gourhan, *Techniques et langage* (1964) 228, 233, et *Le Geste et la parole, La Mémoire, les rythmes* (1965) 157.

48. Perry Anderson, *Les passages de l'Antiquité au féodalisme*, trad. (1977) 238 avec en n.3 un exposé sur le sens du débat (ch.: 'Le frein du nomadisme').

49. A. Leroi-Gourhan, (n.47) 236.

50. J. Sapin, 'La géographie humaine de la Syrie Palestine au IIe millénaire av. J.-C. comme voie de recherche historique', *Journal of the Economic and Social History of the Orient*, 25 (198-) 183; *et* est souligné par l'auteur.

51. Brent D. Shaw, 'Fear and Loathing: The Nomad Menace and Roman Africa', dans C.M. Wells, ed., *L'Afrique romaine*, Les Conférences Vanier (1980) *Revue de l'Université d'Ottawa*, 52 (1982) 25-46. P. Trousset, Notes sur un type d'ouvrage linéaire du limes d'Afrique (n.13).

52. Commode présentation de la question dans le manuel de Y. Garlan, *La guerre dans l'Antiquité* (1972).

53. C.R. Whittaker, 'Land and Labour in North Africa', *Klio*, 60 (1978) 331-362. Brent D. Shaw. (n.46) ainsi que *Pastoralists, peasants and politics: the nomad-sendentarist interaction in North Africa under the Roman Empire*, (thesis Cambridge 1978). Mais ce n'est pas le cas de E.W. Fentress, *Numidia and the roman army*, BAR International Series, 53 (1979).

54. Sur cette question, on pourra se reporter au débat dans *Annales ESC* 33 (1978): Y. Thébert, 'Romanisation et déromanisation en Afrique: Histoire décolonisée ou Histoire inversée?', 64-82; H. Benabou, 'Les Romains ont-ils conquis l'Afrique?', 33-88; Ph. Leveau, 'La situation coloniale de l'Afrique romaine', 89-92.

55. Ph. Leveau, (n.29) 487.

56. *Id.* (n.29) 499.

57. M. Benabou, 'L'Afrique', dans *L'impero romano e le strutture economiche e sociali delle province*, (M. Crawford, dir.) (1986) 127-141.

58. E.W. Fentress, 'Forever Berber?', *Opus* 2 (1983) 161-175.

59. Sur les prospections en Libye, *cf* n.23. En Tunisie, outre les prospections menées par les chercheurs de l'INAA, ainsi celles de H. Slim aux environs de Thysdrus: 'Nouveaux témoignages sur la vie économique à Thysdrus', *BCTH*, 19B (1985) 63-85, il faut noter le grand intérêt pour notre sujet de celles qui sont dirigées par Bruce Hitchner pour la partie américaine, autour de Kasserine depuis 1982, dans le cadre d'un accord entre l'INAA et l'Université de Virginie.

61. Brend Shaw a écrit l'histoire de ce mythe: 'The camel in Roman North Africa and the Sahara: history, biology, and human economy', *Bulletin de l'IFAN*, 41, B, 4 (1979) 663-721. Pour la période préhistorique, on se reportera à G. Camps, 'Origines de la domestication en Afrique du Nord et au Sahara', dans *Le sol, la parole et l'écrit, 2000 ans d'histoire africaine*, (Mélanges en hommage à Raymond Mauny) (1981) 547-560 avec une bibliographie.

62. H. Hurst, P. Paskoff et F. Rakob, 'Position du niveau de la mer et déplacement de la ligne de rivage à Carthage dans l'Antiquité', *C.R. Acad. Sc. Paris*, 300, sèrie 11, n°13 (1983) 617.

MOUNTAIN ECONOMIES IN SOUTHERN EUROPE

Thoughts on the early history, continuity and individuality of Mediterranean upland pastoralism

Introduction

The perspective of this paper is historical rather than ethnological. This is necessarily the case given my area of specialization (antiquity and early middle ages) and the relative paucity of information bearing on Mediterranean mountain societies before documentary evidence becomes available, in the middle ages or later. (This is not to say that medieval documents, consisting typically of the records of large, ecclesiastical landholdings (e.g. Wickham 1982; Rowland 1982), are of the stuff to produce studies comparable with E. Le Roy Ladurie's 'Montaillou'). However, students of antiquity no longer passively accept the deficiencies of the conventional source material, which is for the most part literary and city-oriented, issuing from a civilization centred on the lowlands, coasts and islands of the Mediterranean region. Archaeology, through the utilization of new techniques such as carbon-dating, pollen analysis, zooarchaeology and dendrochronology (Bottema 1974, Kuniholm and Striker 1983, Payne 1985), improved methods of field survey and excavation (Barker et al. 1978, Lloyd and Barker 1981), and, not least, methods, theories and information gleaned from the social and life sciences (cf. Halstead 1984), can now establish a firmer outline of the chronology and changing character of human and environmental development in the Mediterranean. In addition, historians and archaeologists are now learning to use modern or recent quantitative data for climatic behaviour and agricultural performance to construct agroclimatological models of past Mediterranean societies (Garnsey et al. 1984). Also important is the development of ethnoarchaeology, the study of the archaeology created by modern or recent societies, of hunter-gatherers or nomadic or transhumant pastoralists, with a view to identifying and characterizing comparable societies in the past from the archaeological record (Barker, in press). Unfortunately the emergence of ethnoarchaeology is proceeding 'pari passu' with a diminution in quantity and quality of the raw material on which its progress depends – in the case that concerns us, the traditional mountain communities of the Mediterranean, which are fast declining under the impact of economic forces. The need to catch these societies before they disappear should be brought home to anthropologists, as well as to historians and archaeologists. Despite a few successful studies (Campbell 1964, Koster 1977, Ravis-Giordani 1983) this is an underexploited area of scholarship – in contrast with the study of Alpine, and in particular Andean mountain societies (e.g. Netting 1972, 1982, Rhoades and Thompson 1975, Murra 1975, Lehmann 1982), or for that matter, Mediterranean peasant societies (Davis 1973, Forbes 1982). In what follows I pursue certain broad themes relating to the historical development of man's interest in and exploitation of what was and still is, for most inhabitants of the Mediterranean, a peripheral and inhospitable world.

The marginality of the mountains

Strabo, a Greek geographer patronised by the emperor Augustus, wrote with distaste of the mountain peoples of the Iberian peninsula who lived on goat's meat, ate acorn bread for two thirds of the year, drank beer not wine, used butter not olive oil and exchanged by barter (C 154).

Strabo reveals the prejudices – and some of the dietary preferences – of the cultured urban elite of Mediterranean antiquity. In the literature of antiquity mountains are the haunt of the uncivilized and savage, whether men or beasts. Civilization is centred on the 'polis' or 'civitas'; it does not extend much beyond the lowlands, where in the main the permanent communities of the Greco-Roman world were to be found. When the Romans, the supreme conquering nation of the ancient world, extended their advance into the interior of the Iberian peninsula or the backblocks of France, for example, the prerequisite and symbol of their success was the abandonment by the conquered tribes of their hill-top refuges and their resettlement as communities of farmers in the plain, preferably within the territory and juridical and fiscal control of an urban centre.

Goods were exchanged with the inhabitants of the mountains. Strabo knew that the Carretanians on the Celtic side of the Pyrenees gained useful revenues from their excellent hams (C 162), that the Ligurians brought down to Genua flocks, hides, honey and timber, taking back olive oil and Italian wine, their own wine being harsh and flavoured with pitch; their preferred beverages, however, were milk and a drink made of barley (C 202). Strabo no doubt represents a general view when he states that the mountain peoples were forced into exchange relationships with the people of the plain because of the poverty of their own territory, and that their natural instinct was to plunder. The mountains were the home of brigands. In any case there is no hint in Strabo that the inhabitants of the plains were interested in any direct exploitation of the resources of the mountains. We note for example the detail that the Ligurian uplanders pastured their flocks in the coastal districts as well as in the mountains, but Strabo does not identify any plain-based pastoral industry which used the pastures of the Ligurian alps: Strabo, then, knew of the operation in Liguria of 'transhumance inverse', migration of livestock proceeding from a mountain base (a practice of modern Ligurian shepherds, cf. Lamboglia 1923) but not apparently of 'transhumance normale', migration initiating from the plain. Strabo, a most casual collector and dispenser of information, was less than well-informed on this score.

Still, the general impression that Strabo conveys, of the marginality of the mountains from an economic point of view, is largely correct. The Mediterranean peoples have traditionally been cereal consumers on a massive scale; in antiquity perhaps 65-75% of their daily food energy requirements were provided by the annual cereal crop (cf. Foxhall and Forbes 1982). Seed crops can be grown at high elevations, up to perhaps 1500-1800 metres in Southern Europe – not however the preferred cereals of antiquity, the various primitive wheats and barleys. Potatoes and maize were unknown. But yields were low and variable, reflecting the relative unsuitability of the soils and the shortness of the growing season, which decreases as altitude increases – falling for example from around 170 days at 1000 m to around 95 days at 2000 m (Baticle 1974, 83). Early farmers made use of lower slopes for the deep-rooted walnut, chestnut, vine, fig, and to a limited extent, that is, in frost-free zones, the olive. However, insofar as the peoples of the Mediterranean lowlands were interested in the economic exploitation of the mountains in the age before hydroelectricity, tourism and the cult of health (if we leave aside minerals, in any case only sporadically present), it was summer pasture which attracted them. A characteristic feature of livestock raising in the Mediterranean region is transhumance, 'Transhumanz', 'yaylag' pastoralism (a Soviet anthropological term, where 'yaylag' is Turkish for summer highland pasture, Khazanov 1983, 23), defined as "l'oscillation annuelle du bétail entre deux zones de pâturage que séparent des espaces consacrés à des formes d'exploitation différentes" (Arbos 1922). In some specific historical contexts the pastoral industry has established a position of relative independence in relation to the agricultural economy. The mass of the population of the Mediterranean, however, has continued to depend for their subsistence on the products of agriculture. Agriculture in comparison with animal husbandry is both more efficient in furnishing man's calorific requirements, and more productive per unit area, given choice of area, that is, lowlands. A family of six must consume about 115 sheep of carcass weight 20 kg or about 130 sheep of 15 kg if mutton is to be their main source of food energy (Dahl and Hjort 1976, 220); such a family would have to run about 360 live sheep and control a grazing area of about 3.6 km^2 (Halstead 1981, 314; 1984, 7.3). For a largely lowland population of limited land resources, pure pastoralism was no alternative to agriculture as a way of life.

Was there a specifically Mediterranean form of pastoralism?

The best known and most conspicuous examples of Mediterranean transhumance in past history are perhaps the 'Mesta' of Castile, and the 'Dogana della Mena delle Pecore di Puglia'. The 'Mesta' ran for five and one half centuries from its establishment by the court in 1273 and involved at its height in the early sixteenth century the movement of four to five million sheep (Klein 1920). The 'Dogana' governed the movements of sheep (amounting to around five and one half million at its peak in the late seventeenth century) in the Central and Southern Apennines from its inauguration by Alfonso I of Aragon in 1447 until its abolition in 1806 (Di Cicco and Musti 1970-75, Sprengel 1971). In the Balkans it is the transhumant or semi-nomadic 'Vlachs' and 'Sarakatsani' who have attracted most attention among historians and anthropologists (Wiegand 1894, Wace and Thompson 1914, Campbell 1964, Hammond 1976, Winnifrith 1983).

For early evidence of a developed pastoral industry in Italy, we can go back to classical Roman times to Varro, a Roman senator, antiquarian, and composer of three books on *Res rusticae* in the mid-first century BC, that is, a generation before Strabo. Varro knew of rich Romans who owned transhumant flocks of sheep in Italy. Indeed he was one of them, sending a substantial flock, presumably all or most of the 800 sheep that are attributed to him, the distance of around 250 km between the Tavoliere, the great plain of Puglia, and the mountains near Reate east of Rome. The origins of what has been called 'horizontal or Mediterranean' transhumance practised by Varro and his friends and in later ages by Roman emperors is much debated (Skydsgaard 1974, Pasquinucci 1979). Rome's conquest of Samnium had removed the main political barrier to the long-distance movement of livestock in Italy by the early third century BC. But there may not have been any substantial investment in long-distance transhumance for another century, that is to say, until suitable conditions for its development were created with the establishment of peace in Italy following the Carthaginian invasion and occupation of the South, the large-scale confiscation of land from Italian supporters of Hannibal that followed his retreat, and the enrichment of Roman magnates in the wars in Spain, the Eastern Mediterranean and North Africa (Toynbee 1967, Brunt 1971, Frayn 1984). By this account, then, long-distance transhumant pastoralism 'took off' in Italy in the course of the second century BC, a century before the lifetime of Varro.

Some would give Italian long-distance transhumance a longer history. Not Franciosa, who in a lengthy treatment of the subject put forward the view that in the pre-Roman period political conditions, in particular territorial conflicts, made possible only a simple form of livestock raising in the Apennines, 'quella stanziale con movimenti localizzati tra le vallate e i sovrincombenti rilievi montuosi' (Franciosa 1951, 52). A recent, authoritative treatment, however, makes the following generalized statement in support of a pre-Roman origin for long-distance stock movement:

> Tanto più che nella maggior parte delle aree annesse, l'economia pastorale aveva grande importanza e la transumanza doveva essere stata praticata da epoca pre- e protostorica, con spostamenti stagionali di raggio più o meno

esteso, verosimilmente regolati in tempo di pace da accordi e consuetudini che permettevano alle comunità di pianura e di montagna di sfruttare i rispettivi pascoli complementari (Pasquinucci 1979, 93-4).

The evidence on which this statement is based is essentially archaeological: it consists of excavated artefacts and faunal samples and the implications of site locations. The presence of hill-top sites, shrines to Hercules and statuettes of the same god, or for that matter 'fibulae' (clasps) in scattered sites of the Abruzzi and in the approaches to the mountains in proximity to drove roads, does point to their use for the passage of flocks (Barker 1973, 1975, La Regina 1970-71, 1975, di Niro 1977, Patterson 1984). However, it would be unwise to ascribe to pastoralism a continuous, trouble-free history. Franciosa's intuition concerning the unstable political situation in pre-Roman times carries weight. One might add that the defeat of the main Samnite forces, and the colonization of key points in enemy territory such as Isernia and Benevento, are unlikely to have produced tranquillity overnight in the conquered area. Again, caution is advisable in estimating the scale of pastoralism in the context of what was probably an underpopulated society practising a relatively primitive mixed farming economy. A statement such as the following needs to be tested in the light of the demographic and economic setting of the mountain communities concerned and their neighbours:

> Che il carico di bestiame (sc. in età preromana) fosse consistente e richiedesse vaste estensioni pascolive, si può intuire anche dal fatto che lana e pelli erano indispensabili per l'abbigliamento e l'armamento (Pasquinucci 1984).

In the Balkan peninsula also, the early history of transhumant pastoralism is obscure and controversial. Evidence is in fact notoriously thin even for the classical period of Greek history (Georgoudi 1976). This has not prevented scholars from surmising that pastoralism, transhumant or semi-nomadic, has been a constant feature of the Northwest Balkan peninsula throughout history – and prehistory; and even that the present-day shepherds are direct descendants of their presumed predecessors in pre-classical Greece (Hoeg 1925).

Whatever the truth about long-range transhumant pastoralism in the context of pre-Roman Italy, there is no difficulty in admitting the presence in this period of the kind of pastoralism envisaged by Franciosa (above), that is to say, a modest form of transhumance which involved the movement of flocks over short distances from pastures in the plain to others on high ground, or vice versa. It is to be noted that its existence is explicitly accepted by Pasquinucci in the passage quoted above ('con spostamenti stagionali di raggio più o meno esteso'), if only in passing, while elsewhere brief reference is made to 'piccola' as opposed to 'grande' transhumance.

Similarly, for the Roman period, it would be a mistake to assert that long-distance transhumant pastoralism displaced earlier forms of animal husbandry, or that Varro and classical Romans knew and practised no other. For classical as for pre-classical Italy it is safe to postulate the coexistence of several patterns of animal husbandry. The enterprises of Roman aristocrats and emperors (Corbier 1983) are simply the most conspicuous, and it is not surprising that they have monopolised the attention of historians of antiquity, rather as the 'Mesta' and 'Dogana' have caught the eye of students of the medieval and early modern periods at the expense of smaller-scale pastoral movements. In the Biferno valley (Molise), the site of a recent extensive archaeological field survey, villagers at the valley head apparently drove their flocks a few miles into the lower valley in a descent of about 500 m. This same valley in its upper reaches is crossed by major north-south transhumant tracks joining Apulia and the Abruzzi (Barker and Lloyd 1981). The Roman writer Columella, whose *De re rustica* in ten books appeared in the middle of the first century AD, that is, about one hundred years after Varro's work, actually manages to ignore transhumance movements altogether in describing what appears to be entirely a farm-based animal husbandry. Varro himself gives most of his attention to the raising of stock at or near the farm. He does advise the pasturing of cattle 'in wooded land' (*in nemoribus*), adding 'and those which spend the winter along the coast are driven in summer into the leafy mountains' (2.2.8); and he observes that mules, a speciality of his home district Reate, 'are driven into the mountains in summer' (2.8.5). But it is not to be supposed that in either case the destination of the animals was distant rather than local upland ranges. Meanwhile in the case of sheep, Varro contrasts the flocks of the home farm (*uillatici greges*) with those who use the 'glades and woody pastures' (*saltus et siluestres loci*) (2.2.8), self-evidently not a description of high mountains. Similarly, herdsmen are divided into those who work the *calles* (Latin for tracks, 'tratturi'), and those who come back to the farm every day (2.10.1). The *calles* in question need not be the long-distance 'tratturi'.

In sum, there is a case for saying that a form of long-distance transhumance developed in the Mediterranean which was distinctive to the region, and that it evolved, in Italy anyway, when an elite possessed the resources in cash and in land to run sizeable flocks between properties in their possession or under their control. Varro certainly owned land around Reate; presumably it was for this reason that his sheep travelled all the way to the Reatine mountains instead of pasturing for a fee in some equally suitable terrain further south and closer to their base, which was Apulia (2 pref. 6). It does not follow, however, that this was the typical Mediterranean form of pastoralism. To this extent the terminological distinction sometimes employed between 'horizontal or Mediterranean' and 'vertical or Alpine' transhumance is a barrier to understanding. When scholars refer to the former, they evidently have in mind the seasonal movement of livestock, typically sheep but also goats, over long distances. But we have identified another form of stock movement, namely short-range transhumance, which is equally 'Mediterranean' and more accurately described as 'vertical' than 'horizontal'. In any case, 'horizontal' must be shorthand for 'vertical plus horizontal'. However far the flock travels along range-roads to reach the seasonal pastures, winter or summer, that are their destination, the 'raison d'être' of such stock migration is 'vertical' movement between lower and higher ecological zones. To add to the confusion, there are writers who choose not to refer to the 'vertical' movement of livestock in an Alpine setting (monticazione, alpeggio, alpinage) as 'transhumance'. Blache in his classic *L'homme et la montagne* claims that transhumance in France is practised only in the southern part of the Alpine chain and the Pyrenees; the North French Alps and Switzerland are excluded, because there is no migration from plain (or 'Mittelland') to mountains, which are left to the 'montagnards' and inhabitants of the valleys to exploit (Blache 1933). Blache was apparently unaware of a pattern of sheep-raising which established itself in Switzerland subsequent to the First World War, involving the transport of the flock in summer from the Mittelland to the mountains. Gubler-Gross, in his recent dissertation on transhumance in modern Switzerland, focusses almost entirely on this phenomenon, to the exclusion, in the manner of Blache, of the traditional seasonal movement of cattle between valley and lower mountain slope and upland meadow (Gubler-Gross 1962). Yet the Alpine districts were turning away from cereal cultivation towards extensive livestock raising, involving cattle, from the eleventh century, in a development which took five or more centuries to complete (Bergier 1983, 8). There are other apparent illogicalities in the literature of pastoralism; one is the refusal in some quarters to apply the label 'pastoralism' to any system of animal husbandry in which agricultural activities are a not negligible element (Delano Smith 1979, 239). In our case, there is a need to apply a less restrictive definition of transhumance, one which for example takes in short-range as well as long-range stock movements and can therefore embrace animals other than small ruminants.

One by-product of the adoption of a single model with two-dimensional variations, would be the restoration of the concept of 'verticality' to the central position that it must occupy in the analysis of any pastoral system which utilizes mountains. One can then proceed to draw distinctions between Mediterranean and Alpine (and Andean) verticality. The essential difference between Mediterranean and Alpine pastoral movements becomes clear if we ask why they were undertaken. Briefly, Central and Southern Italians, and other Mediterranean peoples, take their sheep to the mountains when, and because, there is no more food for them on the parched plains. Alpine farmers, on the other hand, drive their cattle to the high grasslands in the same, somewhat abbreviated, summer season, in order to free the lower fields for the production of forage crops and hay for consumption by the cattle after their autumn return and stabling. Meanwhile, in parallel with transhumant pasturing, some of the middle and high grasslands are set aside for haymaking, since high altitude grass is thought to be beneficial to the animals. Haymaking is not generally feasible in the Mediterranean uplands.

Climatic determinism – and its limits

Having accepted the coexistence of different forms of pastoralism in the Mediterranean, we should next recognize the possibility of discontinuities in the historical development of pastoralism. The case for continuity is normally put in terms of the climatic and ecological constraints on man's existence and survival in a Mediterranean setting. The argument is a traditional one; it has been accepted without question by generations of scholars. The picture of historical reality which it implies has been challenged and the primacy of social, economic and political factors asserted only recently (Lewthwaite 1981, 1984). In Andean scholarship the converse has occurred: Murra's explanation of verticality in terms of superstructural factors (verticality is seen as essentially a mechanism of control) (Murra 1975) is being countered by rival explanations emphasizing the environmental constraints on production in mountain zones (Guillet 1981, 1983).

The case for environmental constraints on man's operations in the Mediterranean region can be put simply. In the Mediterranean, plant growth is inhibited by cold, especially in the uplands (the lowlands generally experience a mild winter), and by aridity. Spring growth can represent as much as 85% of the growth of the year, as has been demonstrated in stations in the 'Murge' in South Italy and the 'los Pedroches' plateau north of the Sierra Morena in Spain (Baticle 1974, 26). Outside the (untypical) irrigated meadows and marshes, the (typical) summer drought dries up the natural pastures of the lowlands and forces the livestock into the mountains, where the snows have melted to expose fresh pastures. The seasonal movement of stock is confined in large part to the smaller ruminants, expecially sheep, often the goat, occasionally the pig. Sheep and goats are light and mobile, can cope with steep and rough terrain, and can live off the poorer pastures, the 'maquis' and 'garriga', which are common on Mediterranean slopes and uplands.

This, the conventional picture of Mediterranean pastoralism, needs to be refined to make allowance for regional variations in climatic and vegetational conditions, which affect for example the length and severity of the summer drought. Its main weakness, however, is its limited usefulness as a tool to explain the shape that animal husbandry has taken in the Mediterranean throughout history. For example, it would be quite mistaken to seek to explain the phenomenon of the aragonese 'Dogana' or for that matter the quantitatively much less significant enterprises of classical Romans with reference to a supposed iron law of climate and geography. Blache writes: 'Aux origines de la transhumance on trouve moins l'ingéniosité humaine que les lois de la nature vivante' (Blache, 1933). For Franciosa, transhumance is 'una consequenza del clima ... con una causalità puramente geografica' (Franciosa 1951, 11, cfr. 17). This is at best a halftruth – as in fact Franciosa appears to concede later when he states that transhumance represents 'un adattamento costruttivo alle condizioni fisiche ... e non meno a quelle sociali e igieniche (di instabilità, di insicurezza, di malaria)' (p. 51). Transhumance was not 'naturale', 'un fenomeno geografico': it was man-made, the product of social, economic and political conditions specific to particular historical periods.

This is particularly true in the case of extensive pastoralism, or long-range transhumance. One requirement for its institution, or expansion, is the economic and political weakness of local agriculture often following a period of warfare, as in classical Rome and medieval Spain; another is the existence of an outlet for the products of pastoralism, in antiquity and the Middle Ages principally wool, in the modern period more commonly cheese and meat. The size of the industry was determined above all by the strength and stability of market demand. In the last hundred years, the fortunes of Sardinian pastoralism have been closely related to the fluctuating demand for Sardinian cheese in Italy and other countries, including North America; while Roquefort sustains Corsican pastoralism at its relatively high level of significance. The estimates (or assumptions) of scholars of antiquity concerning the size and importance of Italian transhumance in the Republican and Imperial period are invariably exaggerated because they are not based on a realistic assessment of the size of the market, and on an understanding of the economy of extensive pastoralism in general. 'Mesta' and 'Dogana', which responded to a Europe-wide demand for wool for textiles, would have far outstripped the pastoral industry of the classical Roman period.

In the second place, political conditions affected the health of the pastoral industry. The end of the Roman Empire may have ushered in a period of uncertainty for pastoralism in the Apennines (Wickham 1982, 50, 1983, 34), although the organizational capacity of Lombardian or Frankish governments is not to be discounted altogether. In the same period insecurity in Sardinia and Corsica, because it was concentrated in the littoral and lowland areas, had the paradoxical effect of stimulating pastoralism in the less vulnerable interior of those islands in the Early Middle Ages (Lewthwaite 1984). At the other extreme, 'Mesta' and 'Dogana' were implemented by powerful political authorities, which, spurred on by the prospect of lucrative tax revenues, were able to control the economy of extensive regions in Spain and Italy, respectively. The parallel with Andean verticality as interpreted by Murra does not need stressing (Murra 1975, D'Altroy and Earle 1985, 189). In classical Rome, pastoral enterprises were regulated but not institutionalized: a less elaborate apparatus of control and a less dirigist ideology combined with strictly limited marketing possibilities to produce a smaller scale pastoral industry.

The creation of mountain pasture

We have been considering the role that man has himself played in shaping the historical development of pastoralism. But this has an aspect thus far untreated. The physical environment imposes constraints on human behaviour. Yet man has shown himself capable of transforming his natural surroundings. In particular, the geographical conditions under which long-distance transhumance could flourish are in some important respects man-made.

Transhumance by definition involves stock movements between two sets of seasonal pastures, in the plain and in the mountains. But we need to ask what pasture was available in the mountains.

Natural grassland is relatively rare in the Mediterranean. This is not because the mountains are everywhere too low. It is true that we are dealing with mountains that rarely top 3000 m, mere anthills from the perspective of an Andean or Himalayan geographer. However, the lack of height is offset by the phenomenon of 'Massenerhebung', the compression of zones of climate and vegetation on small mountains relative to altitude, compared with the interior of big mountain ranges. The natural upper limit of woodland in South Greek mountains, for example, is about 1800-1900 m, as is indicated by relict trees in cliffs. Thus on Parnassus and Cretan Ida a zone of 500-600 m separates treeline from (more or less permanent) snowline. If natural grassland is rare in Greece, the reason is not so much the lack of altitude as the hard limestone and marble rock which inhibits the formation of suitable soils (O. Rackham, pers. comm.).

However, man has lowered the treeline and created an artificial, pseudo-alpine meadow. The pastures thus formed are frequently low in mineral and protein content and also very liable to the summer drought, which, in the traditional model of transhumant pastoralism, is a feature of the lowlands (Baticle 1974). The irregularity of the mountain climate in Sardinia means that insofar as transhumance takes place (and most sheep are not transhumant), the flocks are driven not up to escape the summer drought but down to avoid the winter cold (Le Lannou 1941).

The deforestation which created the middle-altitude pastures is usually attributed to organized pastoralism. 'In Spain after the Reconquest, deforestation became the accepted policy of the pastoralists, especially after 1273 when the Mesta of Castile was established, first along the drove roads, and then in the pastures ... With their Spanish experience, the Aragonese did the same thing in Southern Italy after 1300' (Houston 1954, 114). 'Trees were not tolerated on or near the grazing lands because, it was held, their roots absorbed too much moisture from the soil and so reduced the grass cover. Whole forests were cut down by or as a result of Dogana pressures' (Delano Smith 1979, 246). Clearance of high woodland is thought to have accelerated markedly in the nineteenth century. We read that it was in this period that half the woods that covered the slopes of the Gran Sasso and other high mountains in the Abruzzi were destroyed (Franciosa 1951, 71).

Was antiquity too a period of significant tree-clearance in the mountains? Woods can be cut down (or burned) more than once; they come back when pastoralism is in retreat. For this reason wildwood is on the increase in the Mediterranean mountains today. It is thus in theory possible that deforestation was carried out to the same extent in Italy in the Roman period as in the High Middle Ages – although the probability of this seems slim in view of the relatively small scale of Roman pastoralism. In the Balkans the problem is rather different: we know that Greece was well-wooded in antiquity, especially in the Pindus range in the Northwest – this is confirmed by the palynological evidence, where it is available in sufficient quantity and quality (Bottema 1974; Halstead 1981, 1984). And as we saw we have virtually no information about uplands pastoralism in ancient Greece. But it would be rash to state that woodland cover in antiquity was such as to preclude transhumance altogether. This means among other things that the continuous existence of pastoral transhumance in the Greek mountains cannot be ruled out, even if we are not inclined to accept particular continuity theories, such as those that find the origins of today's 'Sarakatsani' in antiquity (Hoeg 1925), or ascribe the emergence of Doric and Northwest Greek dialects to the infiltration into settled early archaic Greek communities of transhumant or nomadic shepherds established in the Greek mountains (Kirsten 1983).

Conclusion

Geographical facts do not account for long-range transhumance, or even for the existence of a pastoral industry at all. They rather govern the form, or more accurately, forms, that pastoralism can take in a Mediterranean setting, whether as a partner with agriculture and arboriculture in the mixed farm characteristic of the traditional peasant economy, or as a semi-independent, relatively specialized industry operating over an extended area, or as something in between. Of these two 'pure' forms the first has strong grounds for being considered indigenous to the Mediterranean region and a more or less permanent feature of the landscape. In it the pastoral element is underdeveloped and engaged in a symbiotic association with the other sectors. It is also mixed in character, involving bovines for traction, swine for meat and ovicaprines for clothing and milk products – not normally for meat, for sheep are less efficient converters of plant food into meat than pigs and have other uses. The fodder shortage of the summer months when vegetational growth ceases varies in scale and severity with local climatic conditions. It is met, or alleviated, by the combination of several remedies. The first is the use of stubble, fallow and wasteland for grazing, typically by sheep, which in the process return manure to the soil. Sheep manure is both qualitatively superior and ample – a modern sheep is capable of producing 500 kg of manure p.a. (Baticle 1974, 34) – and in some contexts has been regarded as the sheep's main asset. The second food resource for animals in the lean season is arboreal. An observer of Umbrian agriculture at the turn of the eighteenth century describes the careful exploitation by the peasants of foliage as well as all kinds of vegetable growth:

> Le paysan met à profit toute l'herbe qu'il peut faucher sur le bord des fossées, toute les feuilles qu'il peut arracher à la vigne ou aux peupliers, toutes les plantes parasites qu'il peut trier entre ses blés et toute la paille qu'il peut hacher et mêler aux fourrages vertes; il va souvent jusqu'à dépouiller en automne les arbres fruitiers de leurs feuilles, et même les figuiers, malgré l'acreté de leur sève; il réussit enfin par cet esprit d'économie, s'il a peu de bêtes en proportion de son terrain, à en avoir beaucoup et beaucoup de fumier en proportion de son fourrage (Desplanques 1969, 419).

But we should also bear in mind traditional forms of woodland exploitation, in particular the acorn-hog economy of Southwest Iberia or Umbria, or the chestnut-hog economy of the Tyrrhenian islands (Parsons 1962, Desplanques 1969, 420, Lewthwaite 1982), Umbria, the Tyrrhenian islands, and most notably the South West region of the Iberian peninsula (Desplanques 1969 420, Lewthwaite 1982, Parsons 1962). Thirdly, the broken nature of much of the landscape, particularly in the Balkan and Italian peninsulas, makes the local movement of livestock between different ecological zones desirable and feasible. Long-range transhumance in contrast is inconvenient, expensive and detrimental to agriculture in that it deprives the lowland soils of a valuable fertilizer. Its emergence is associated with historical change, whether social and demographic, or in respect of the shape or structure of political authority, or in the level and nature of consumer demand for its products. Similar factors have engineered its recent rapid decline and will set the limits of its future development.

Note: In writing this paper I have benefited in particular from the assistance of Robin Donkin, Paul Halstead, James Lewthwaite, John Patterson, Oliver Rackham and Jonathan Thompson. I am grateful also to Graeme Barker and Chris Wickham for making work available in advance of publication, and to Carmine Ampolo for sending me material unavailable in the United Kingdom.

JESUS COLLEGE, CAMBRIDGE PETER GARNSEY

BIBLIOGRAPHY

Arbos, Ph., *La vie pastorale dans les Alpes françaises* (1922).

Barker, G., 'Cultural and economic change in the prehistory of central Italy', in: C. Renfrew (ed.), *The explanation of cultural change: models in prehistory* (1973) 359-70.

Barker, G., 'Prehistoric territories and economies in central Italy', in: E.S. Higgs (ed.), *Palaeoeconomy* (1975) 111-75.

Barker, G., 'Can we recognize transhumance in the archeological record?', in press.

Barker, G., Lloyd, J., Webley, D., 'A classical landscape in Molise', *Papers of the British School at Rome*, 46: 35-51, 1978.

Baticle, Y., *L'élevage ovin dans les pays européens de la Méditerranée occidentale* (1974).

Bergier, J.-F., *Histoire et aménagement du territoire: De quelques expériences de la Suisse médiévale*, Disp. Nr. 56, Dokumente und Information zur Schweiz. Orts-, Regional- und Landesplanung. Sondernummer: Geschichte der Landesplanung: 6-13, 1980.

Blache, J., *L'homme et la montagne* (1933).

Brunt, P.A., *Italian manpower 225 BC-AD 14* (1971).

Bottema, S., *Late quaternary vegetation history of North-western Greece* (1974).

Campbell, J.K., *Honour, family and patronage. A study of institutions and moral values in a Greek mountain community* (1964).

Corbier, M., 'Fiscus and Patrimonium: The inscription and transhumance in the Abruzzi', *Journal of Roman Studies* 73: 126-31, 1983.

Dahl, G., Hjort, A., *Having herds* (1976).

D'Altroy, T.N., Earle, T.K., 'Staple finance, wealth finance, and storage in the Inka Political Economy', *Current Anthropology* 26: 187-206, 1985.

Davis, J., *Land and family in a South Italian town* (1973).

Delano Smith, C., *Western Mediterranean Europe* (1979).

Desplanques, H., *Campagnes ombriennes: Contribution à l'étude des paysages ruraux en Italie centrale* (1969).

Di Niro, A., *Il culto di Ercole tra i Sanniti Pentri e Frentani: Nuove testimonianze* (1977).

Forbes, H.A., *Strategy and soils: Technology, production and environment in the peninsula of Methana, Greece*. Diss. Penn. 1982.

Foxhall, L., Forbes, H.A., 'Sitometreia: The Role of Grain as a Staple Food in Classical Antiquity', *Chiron* 12: 41-90, 1982.

Franciosa, L., 'La transumanza nell' Appennino centro-meridionale', *Mem. Geog. Econ.* (Centro C.N.R. St. Georg. Econ. Napoli) 3:7-97, 1951.

Frayn, J., *Sheep-rearing and the Wool Trade in Italy during the Roman period* (1984).

Garnsey, P., Gallant, T., Rathbone, D., 'Thessaly and the grain supply of Rome in the second century B.C.', *Journal of Roman Studies* 74: 30-44, 1984.

Georgoudi, S., 'Quelques problèmes de la transhumance dans la Grèce ancienne', *Revue des études grecques* 87: 155-81, 1976.

Gubler-Gross, R., *Moderne Transhumanz in der Schweiz*. Diss. (1962).

Guillet, D., 'Land tenure, ecological zone and agricultural regime in the Central Andes', *American Ethnologist* 8: 139-56, 1981.

Guillet, D., 'Toward a cultural ecology of mountains: The central Andes and the Himalayas compared'. *Current Anthropology* 24: 561-74, 1983.

Halstead, P., 'Counting sheep in Neolithic and Bronze Age Greece', in: Hodder, I., Isaac, G., Hammond, N. (eds.), *Pattern of the past: studies in honour of David Clarke* (1981) 307-39.

Halstead, P., *Strategies for survival: an ecological approach to social and economic change in the early farming communities of Thessaly, N. Greece*, Ph.D. Cambridge 1984.

Hammond, N.G.L., *Migrations and invasions in Greece and adjacent areas* (1976).

Hoeg, C., *Les Saracatsanes* (1925).

Houston, J.M., *The Western Mediterranean World* (1964).

Khazanov, A.M., *Nomads and the outside World* (1983).

Kirsten, E., 'Gebirgshirtentum und Sesshaftigkeit – die Bedeutung der Dark Ages für die griechische Staatenwelt' in: *Doris u. Sparta, Oest. Akad. der Wiss. Phil.-hist. Kl. Sitzungsberichte* Bd. 418. Akten des Symposions von Stift Zwettl 1.-4. Okt. 1980, ed. S. Deger-Jalkotzy (1983).

Klein, J., *The Mesta: a study of Spanish economic history* (1920).

Koster, H., *The ecology of pastoralism in relation to changing patterns of land use in the North East Peloponnese*. Ph.D. Pennsylvania 1977.

Kuniholm, P.I., Striker, C.L., 'Dendrochronological Investigations in the Aegean and Neighbouring Regions 1977-1982', *Journal of Field Archaeology* 10: 411-20, 1983.

Lamboglia, C., 'La transumanza nelli Alpi Ligure', *Atti VIII Congr. Geogr. Ital.*, Pt.2 421-7, 1921.
La Regina, A., 'I territori Sabellici e Sannitici', *Dialoghi di Archeologia* 4-5: 443-59, 1970-71.
La Regina, A., 'Centri fortificati preromani nei territori Sabellici dell'Italia centrale adriatica', *Posebna Izdania* 24: 271-84, 1975.
Lehmann, D., *Ecology and exchange in the Andes* (1982).
Le Lannou, M., *Pâtres et paysans de la Sardaigne* (1941).
Lewthwaite, J., 'Plain tails from the hills: Transhumance in Mediterranean Archaeology', in: A. Sheridan, G. Bailey (eds.), *Economic Archaeology*, B.A.R. International Series 96: 57-66, 1981.
Lewthwaite, J., 'Acorns for the ancestors: the prehistoric exploitation of woodland in the West Mediterranean', in M. Bell, S. Limbrey (eds.) *Archaeological aspects of woodland ecology*, B.A.R. International Series 146: 217-30, 1982.
Lewthwaite, J., 'Pastore, Padrone: The social dimensions of pastoralism in prenuragic Sardinia', in: W.H. Walden, R. Chapman, J. Lewthwaite, R.C. Kennard (eds.), B.A.R. International Series 229 (i): 251-68, 1984.
Lloyd, J., Barker, G., 'Rural settlement in Roman Molise: problems of archaeological survey', in: G. Barker, R. Hodges (eds.), *Archaeology and Italian society: Prehistoric, Roman and Medieval studies*, B.A.R. International Series 102: 289-304, 1981.
Murra, J.V., 'El "control vertical" de un maximo de pisos ecologicos en la economia de las sociedades andinas', *Visita de la provincia de Leon de Huanuco en 1562*, Inigo Ortiz de Zuniga, visitador vol. 2: (1972) 429-76.
Murra, J.V., *Formaciones economicas y politicas andinas* (1975).
Netting, R.Mc.C., 'Of men and meadows: strategies of Alpine land use', *Anthropological Quarterly* 45: 132-44, 1972.
Netting, R.Mc.C., *Balancing on an alp: Ecological change and continuity in a Swiss mountain community* (1982).
Orlove, B.S., 'Ecological anthropology', *Annual Review of Anthropology* 9: 235-73, 1980.
Parsons, J.J., 'The acorn-hog economy of the oak woodlands of Southwestern Spain', *Geog. Rev.* 52: 211-235, 1962.
Pasquinucci, M., 'La transhumanza nell'Italia Romana', in: E. Gabba, M. Pasquinucci, *Strutture Agrarie e allevamento transumante nell'Italia Romana* (III-I sec. A.C. (1979) 79 182.
Pasquinucci, M., 'Aspetti dell'allevamento transumante nell' Italia centro-meridionale adriatica', in: *Sannio: Pentri e Frentani dal VI al I sec. A.C. Atti del Convegno 10-11 Novembre 1980* (1984).
Patterson, J.R., 'The Upper Volturno valley in Roman times', in: R. Hodges, J. Mitchell (eds.), *San Vicenzo al Volturno: the archaeology, art and territory of an early medieval monastery*, B A R, International Series 252: 213-26, 1985.
Payne, S., 'Zoo-archaeology in Greece: a reader's guide', in: N.C. Wilkie, W.D.E. Coulson (eds.), *Contributions to Aegean Archaeology* (1985).
Quézel, P., 'La végétation des hauts sommets du Pinde et de l'Olympe de Thessalie', *Vegetatio* 14: 127-227, 1967.
Ravis-Giordani, G., *Bergers corses: les communautés villageoises du Niolu* (1983).
Rhoades, R.E., Thompson, S.I., 'Adaptive strategies in alpine environments. Beyond ecological particularism', *American Ethnologist* 2: 535-51, 1975.
Rowland, R.J., 'The Sardinian Condaghi: neglected evidence for medieval sex ratios', *Florilegium* 4: 117-22, 1982.
Sprengel, U., *Die Wanderherdenwirtschaft im mittel- und südostitalienischen Raum* (1971).
Toynbee, A.J., *Hannibal's legacy: the Hannibalic war's effects on Roman life*, Vol. II (1965) 286-95.
Wace, A.J.B., Thompson, M.S., *The Nomads of the Balkans* (1914).
Wickham, C., *Studi sulla Società degli Appennini nell'Alto Medievo: Contadini, signori e insediamento nel territorio di Valva (Sulmona)*. Università degli Studi di Bologna. Quaderni del Centro Studi Sorelle Clarke (1982).
Wickham, C., 'Pastoralism and Underdevelopment in the early Middle Ages', *Settimane di Studio* 31: 3-53, 1983.
Wiegand, G., *Die Aromunen* (1894).
Winnifrith, T., 'Greeks and Romans', in: T. Winnifrith, P. Murray (eds.), *Greece Old and New* (1983) 65-94.

ADDING IT UP

The comments that I should like to make concern the relative importance of animal husbandry in the context of the rest of the ancient economy. Even though we now know considerably more about animal husbandry, this does not necessarily increase its importance for the ancient economy. In fact, I want to argue that we now know better that it was not very important. The argument that I want to bring forward consists of four related propositions. Space does not permit more than fairly outrageous simplification; I can only hope that this is somewhat offset by increased clarity.[1]

I. Biologically, the conversion of plant material into meat or milk is very inefficient.[2] Therefore, animal husbandry will provide fewer calories per hectare than for example cereal agriculture. Comparative evidence shows that above certain levels of population density, it is no longer feasible to have substantial numbers of larger animals.[3] The less densely populated parts of the ancient world may have had more animal husbandry than the more densely populated parts. Population density and animal husbandry should be negatively correlated. I believe that this is largely born out by the ancient evidence. Many of the examples of economies with a large pastoral element seem to come from the more peripheral and less densely populated areas.

II. If we turn from production to consumption, we have another possibility of estimating the importance of animal husbandry. Ancient evidence on the composition of the diet of the mass of the population is not abundant, but enough to seem certain that calories from animal sources were only a minor part of the diet in the densely populated core areas of the ancient world. Foxhall and Forbes estimate that cereals accounted for about three quarters of the diet.[4] This is plausible because the cost of food must have been a large part of the budget of the poor; cereals are relatively cheap.[5] Famines in the ancient world are described as shortages of cereals, which again serves to emphasize the latter's importance for subsistence.

Interesting in respect of the difference between core and periphery is the contrast that is drawn by Georges Duby between Roman and Germanic eating habits.[6] The Germanic diet consisted of far larger proportions of meat, dairy produce, garden vegetables, fish and gathered fruit. This makes for a more attractive diet, but it was also one that required far more space.[7]

III. Many contributions have argued against the importance of specialized intensive animal husbandry. This is important because such intensive animal husbandry would have coincided with a high population density, and has as such existed in few European countries (e.g. Holland) in the period prior to the Industrial Revolution.[8] It then provides an example of major animal husbandry in an advanced, rather than a backward context. I can see two reasons for the lack of such specialized intensive animal husbandry in antiquity:

The first reason is that the standard of living of the mass of the population was too low, and income inequality was too large to provide a sizeable market.

The second reason is that trade in subsistence food was insufficiently developed to guarantee a steady supply at reasonable cost to specialized farmers.

IV. The preceding propositions do not exclude the presence of modest numbers of especially small animals, to supplement the diet and provide some secondary products. I was disappointed, however, that the role of working animals, and particularly oxen, has not been raised in any of the papers. I admit that it falls outside what one might call pastoralism, but it is an important issue, and will show related problems. Again, the key is that oxen compete with men for scarce land.

Where population density is high, production per hectare has to be high to feed the people. This usually means a considerably increased labour input per hectare. Unfortunately the law of declining marginal returns will take its toll, and labour productivity will diminish. Labour, however, is not just any factor of production, but consists of humans whose standard of living will inevitably deteriorate. Using more capital, such as an ox, might be a way to increase production without a declining standard of living. But this is neither a socially neutral solution, nor is it ecologically easy in the Mediterranean.

That this must have been a problem is strikingly revealed in early-modern data from Campania.[9] They show that with a plough and oxen the plot that can be worked by a peasant family increases about threefold: the peasant with these is much better off than the peasant who has to do without them. The problem, however, is to find the fodder for the ox. Southern Italian pasture is scarce and of bad quality compared to northern Europe. To quote Gérard Delille: 'Small properties could not feed large animals, and working the land with the hoe, the peasant could not cultivate a plot larger than one that was barely sufficient for his survival.'[10] Once again, the catch of the situation is that the animals compete with people for scarce resources. A small peasant could increase his labour productivity by using work animals, but would thereby make his family's labour redundant. A decent standard of living and a high population density are mutually exclusive in this case. To have enough land to own a plough and oxen must therefore have been one of the most distinctive elements of social differentiation within the peasantry.

Man's tragedy is that his most favoured food, meat, competes with himself for resources. If there are too many people, they have to resign themselves to a vegetarian life. If there are yet more people, the poor among them will have to do without work animals. The classical world achieved much, but one of the prices that it had to pay for this was that many of its inhabitants had to forego most of the pleasant fruits of animal husbandry.

UNIVERSITY OF ROTTERDAM WILLEM JONGMAN

NOTES

1. In my *The economy and society of Pompeii* (1988) I discuss and document more fully some of the concepts used here.

2. David Grigg, *The dynamics of agricultural change* (1982) 68-71.

3. Ester Boserup, *Population and technology* (1981) 15-23.

4. Lin Foxhall and H.A. Forbes, '*Sitometreia*: the role of grain as a staple food in classical antiquity', *Chiron* 12 (1982) 41-90.

5. Carlo M. Cipolla, *Before the industrial revolution. European society and economy, 1000-1700* 2nd. ed. (1981) 28. The remaining quarter or so of food consumption could consist of wine, olive oil and perhaps – I admit – a little bit of meat.

6. Georges Duby, *Guerriers et paysans* (1973) 25-33.

7. I was struck by Bökönyi's – above, p.171-76 – high level of beef consumption in Gorsium. I do not think that such a result could be found for an Italian site. In Jongman (1988) I argue that a diet that consisted largely of cereals is compatible with accepted views of population densities in early Imperial Italy and with the 'traditional' view that two field dry farming was predominant.

8. Jan de Vries, *The Dutch rural economy in the golden age, 1500-1700* (1974).

9. Gérard Delille, *Agricoltura e demografia nel regno di Napoli nei secoli XVIII e XIX* (1977).

10. Delille (1977) 135.

PASTORALISM AND TRANSHUMANCE IN ROMAN ITALY

For the past two years I have been conducting research at Cambridge for a thesis on pastoralism and transhumance in Roman Italy, with the main emphasis on sheep. I am interested in what system or systems of livestock-raising were followed and their place in the economy.

I do not believe in defining the types of pastoralism as a dichotomy between large-scale transhumance involving hundreds or thousands of animals and peasant farmers keeping just a handful of animals. I believe that there are many different sizes of flock attested at different times, fed with various foods, and kept for diverse products, varying proportions of which were sold or exchanged.

Many modern historians, most famously Toynbee,[1] have argued that in the last two centuries B.C. Italy, and in particular the South of Italy, experienced a strong growth in pastoralism and a decline in agriculture. The few references by agronomists to such a phenomenon[2] allow us to believe that some increase in pastoralism did take place, but they do not tell us how much. The best evidence for transhumance is to be found in Varro, but his only specific reference to numbers of animals describes quite small flocks.[3] Cato and Columella prescribe an integration of pastoralism and agriculture.[4] Pliny and Virgil write that you should only practise fallowing if you have enough land, otherwise you should grow fodders in rotation with cereals.[5] It would then be logical to keep animals on the farm to eat the fodder, as Columella tells us,[6] not transhuming to uncultivated areas.

Archaeology has done little to support the belief that pastoralism dominated all of Southern Italy. *Villa* sites such as Monte Irsi[7] where excavators have looked for botanical evidence have shown that fodders were being cultivated, though this could be only for plough oxen. At Metaponto the population does appear to have collapsed in the Republican era and the territory used for pasture,[8] but elsewhere field surveys have indicated different settlement patterns. Even in the Tavoliere we find in the first and second centuries B.C. extensive centuriation, widespread vine and olive cultivation and the export of grain, oil and wine from Apulia.[9]

Since there is little direct evidence for the scale of sheep-raising in Roman Italy, we need to study the situation in other periods for which there is better information and seek to build a model. Ethnoarchaeology can tell us a lot about techniques followed by shepherds and the identification of pastoral sites, but it does not help with broader questions such as what makes pastoralism more important in some periods and regions than in others. Medieval and modern history and social anthropology are more informative.

Professor Gabba has compared Roman 'pastorizia industrializzata' with the Dogana of Medieval Apulia. The Dogana required government control and constant intervention in order to reserve land as winter pasture for transhumant animals. The government prevented farmers from converting the pastures to arable when the grain price was high and forced shepherds to spend the winter in Apulia.[10] The Roman government seems to have been more laissez-faire and there is no reason to believe that it controlled the use of land in this way.

It is therefore more useful to look for a model in a society where the rural economy has been less controlled, somewhere like Sardinia. During the twentieth century there has been some long-distance transhumance of sheep, but this has been only one of many options followed by pastoralists. Others have followed short-distance transhumance which involves a smaller change in altitude, have kept their flocks in the same village throughout the year or have moved animals several times a year in search of different food resources. Moreover, there have been many patterns of ownership and management of flocks, ranging from a few big owners hiring employees, through numerous types of associations of several shepherds to family businesses.[11] In Italy, too, since the abolition of the Dogana di Foggia and other regulatory bodies, diverse systems of feeding sheep have been followed.[12]

These studies have convinced me that in Italy long-distance transhumance would not become the dominant type of pastoralism unless a government wished it to be. I do not believe that the Roman government interfered in the rural economy in this way. Therefore the situation in Roman Italy probably resembled Sardinia in the 1930s more than it did sixteenth-century Puglia. Since a large part of the Italian population were subsistence-farming peasants, their livestock supplied their own and local needs and had little involvement in the cash economy. Those people who owned enough animals to produce a large surplus for sale could conduct long-distance transhumance, but the majority probably followed 'pastorizia stanziale' if the agronomists are a reliable indicator. The situation was far more complex and interesting than it at first appears.

CHURCHILL COLLEGE, JONATHAN THOMPSON
CAMBRIDGE

NOTES

1. A.J.R. Toynbee, *Hannibal's legacy* II (1965).

2. E.g. Varro, *Res rusticae* 2.*praef*.4; Columella, 1.3.12.

3. *R.R.* 2.10.11.

4. E.g. Cato, *De agricultura* 10; Columella, 6.*praef.*

5. Virgil, *Georgica* 1.79-83; Pliny, *N.H.* 18.187.

6. 6.*praef.*2.

7. A.M. Small (ed.), *Monte Irsi* BARS 20 (1977).

8. J.C. Carter in C. Malone and S. Stoddart (eds.), *Papers in Italian archaeology* IV, i BARS 243 (1985) 283-312.

9. G.D.B. Jones, 'Il Tavoliere romano', *Archeologia Classica* (1981); Varro, *Res rusticae* 2.6.5.

10. For a basic account of the Dogana see D. Musto, *La Regia Dogana della Mena delle Pecore di Puglia*, Quaderni della rassegna degli 'archivi di stato' 28 (1964).

11. M. Le Lannou, *Patres et paysans de la Sardaigne* (1941) gives a general picture. There are also many recent local studies, such as B. Meloni, *Famiglie di pastori* (1984).

12. E.g. M. Ortolani, *Il massiccio del Gran Sasso d'Italia*, (1942) 95-105; J. Demangeot, 'Evolution comparée de deux villages Abruzzais', *Revue de géographie alpine* 37 (1949) 671-689; H. Desplanques, *Campagnes ombriennes* (1969) 417-437.

INTERVENTION

Pour le monde romain, il vaudrait la peine de procéder à un inventaire des divers types d'activités pastorales. Nous pouvons distinguer trois niveaux:

– l'élevage intégré à l'agriculture – en Italie, l'élevage pratiqué sur le *fundus*, et recommandé par les agronomes latins (Caton, Varron, Columelle);

– un secteur pastoral d'une agriculture sédentaire, représenté le plus souvent par la transhumance (et lié dans les textes latins à l'évocation du *saltus*), dont la seule description précise concerne l'Italie avec le livre II et quelques passages du livre III des *Res rusticae* de Varron;

– de véritables sociétés de pasteurs, généralement nomades, qu'il faudrait chercher à repérer à l'intérieur ou aux marges de l'empire.

Au prix d'une certaine simplification, on observera que leur étude respective relève souvent de trois démarches différentes: celle de l'archéologue, de l'historien, de l'ethnologue.

Par la photographie aérienne (la Picardie de Roger Agache), la prospection systématique (la Beauce d'Alain Ferdière), elle-même associée à des sondages et des fouilles (l'Etrurie méridionale, telle que nous la présente maintenant Tim Potter), la fouille minutieuse de la *pars rustica* et non plus de la seule *pars urbana* de la *villa* (l'équipe d'Andrea Carandini à Settefinestre, dans l'*ager Cosanus*), avec pour corollaire l'analyse des restes d'animaux et de végétaux retrouvés dans les différents secteurs de la fouille, les méthodes d'approche se sont diversifiées. Taille et spécialisation des bâtiments, rotation des cultures et temps de jachère, fumure du sol, éventail des races animales et des plantes cultivées: les multiples informations sur l'élevage fournies par l'archéologie renouvellent notre connaissance des exploitations rurales, et de l'organisation de l'agriculture.

Les sociétés de pasteurs – le troisième secteur – suscitent une démarche ethnologique de la part des historiens (Brent Shaw) comme des archéologues (Pol Trousset). Elles se signalent par les différences multiples qui les séparent des agriculteurs sédentaires. Mais les sources directes restent rares. Ce qui explique que les représentations du pasteur nomade ('mangeur de viande', 'buveur de lait') restent mieux connues que les sociétés elles-mêmes. Rien de comparable pour le monde romain, et pour cause, aux travaux sur les Valaques des Balkans au XIXe siècle. Autre déséquilibre, géographique cette fois, dans l'information: de la Tingitane au Proche-Orient, ce sont les pasteurs de la zone subdésertique qui ont inspiré le plus d'études, notamment dans le cadre de recherches systématiques sur le *limes*.

Même si, dans la Gaule romaine, il y a eu sans doute plus de moutons sur les grandes fermes de Picardie, et peut-être de Beauce, que de moutons transhumants dans la Crau (Pline l'Ancien), la transhumance fascine les historiens, et l'on comprend pourquoi; elle met en scène le pouvoir, voire l'Etat, elle implique une circulation régulière, souvent à longue distance, des animaux et des hommes entre des espaces contrôlés par des sociétés différentes, elle pose des problèmes d'origine, de continuités (ou de ruptures), et de comparaisons avec d'autres systèmes historiques plus proches de nous, à la fois bien documentés et étudiés.

Des deux grandes organisations transhumantes de l'époque moderne, la *Mesta* et la douane des Pouilles (*Dogana delle Pecore*, dite aussi *di Foggia*), la première ne paraît pas avoir d'antécédents romains attestés, malgré la réputation dont jouissait la laine de Bétique. Des Abruzzes au Tavoliere, et même plus au sud, au territoire de Metaponte, les déplacements de troupeaux se développent au contraire dès la fin de la République, et peut-être avant, selon des itinéraires (*calles*) et sur des distances comparables à ce que nous en connaissons à partir du milieu du XVe siècle. Ils sont de même protégés et surveillés par l'Etat, et soumis à une fiscalité spécifique. Ce qui ne veut pas dire, ni qu'ils aient atteint une même ampleur (plus d'un million de moutons au XVIe siècle), ni que tous les autres éléments économiques et sociaux du système aient été en tous points identiques. L'origine des troupeaux (qui appartiennent à la fin du Moyen-Age aux habitants des Abruzzes), la spécialisation pour la laine plus que pour la viande ou le cuir, en fonction des courants d'exportation à longue distance, les droits de propriété sur les pâturages des montagnes comme sur ceux du Tavoliere, l'organisation de l'agriculture en Apulie, etc, toutes ces questions ont pu avoir à l'époque romaine des réponses différentes, et dans un contexte politique et économique lui aussi très différent.

L'archéologie du territoire peut apporter aussi de précieux éléments d'information, comme le montrent les recherches de Graeme Barker et John Lloyd sur le Molise et le 'survey' du Tavoliere des Pouilles (avec ici le recours possible à toutes les techniques à la fois: photographie aérienne, prospection, fouille...).

Resterait enfin à relier ces analyses sur les différents types d'élevage à ce que nous savons des circuits commerciaux des différents produits (laine, cuirs, fromages, viande...). Un bon exemple en est l'approvisionnement en viande – une viande qui voyage sur pied – des grands marchés de consommation, et d'abord de Rome. Car il permet de mettre en cause d'autres déplacements de troupeaux, cette fois-ci à destination des villes, et signalant d'autres origines: si le cas des oies des Morins, rapporté par Pline l'Ancien, paraît relever de la fable, celui des porcs venus de Campanie ou de Lucanie, qui maigrissent en cours de route, nous rapproche de réalités familières à des époques plus proches de nous.

Sur ces deux derniers aspects, j'aurai l'occasion de revenir prochainement: sur la transhumance des Abruzzes vers les Pouilles que j'ai déjà étudiée (*JRS* 73

(1983) 126-131) dans le cadre d'une communication présentée à Canosa, et sur la consommation de viande à Rome avec un article destiné au volume d' 'Hommage à Pierre Lévêque'.

C.N.R.S. – PARIS MIREILLE CORBIER